The STEM Pathway and Student Retention

Carita Harrell • David G. Capco

The STEM Pathway and Student Retention

Lessons Applied and Best Practices through Peer Mentoring

Springer

Carita Harrell
School of Human Evolution
and Social Change
Arizona State University
Tempe, AZ, USA

David G. Capco
Cell and Molecular Biosciences, School
of Life Sciences
Arizona State University
Tempe, AZ, USA

ISBN 978-3-030-65863-2 ISBN 978-3-030-65865-6 (eBook)
https://doi.org/10.1007/978-3-030-65865-6

This Springer imprint is published by the registered company Springer Nature Switzerland AG
The registered company address is: Gewerbestrasse 11, 6330 Cham, Switzerland

Abstract

This work will introduce methods that will aid in freshman retention (in the transition from high school and to remain in the university of origin) and orient them toward a successful career in STEM. Specific examples of successful approaches will be given as well as detailed plans for how to engage these students. Pitfalls as well as successes will be described. In addition, this work will address the needs of underrepresented populations, including women, ethnic minorities, and students from low socioeconomic status in the STEM fields, which at times have been referred to as "at risk" populations. This work will provide a detailed description of how to develop the students into a cohort that exhibits comradery. Three types of cohorts form—those within the freshman class, those among the upperclassmen, and those between the freshmen and upperclassmen. The program works because the social reality is that the peer mentor has a better repertoire with the first-semester freshmen than the faculty or staff and assists with student success. Factors such as financial aid, policy, and support systems influence student success. In the STEM-related courses, students often struggle with the content and adjusting to the college experience. Research states that a mentorship program supports retention as well as enhances the student experience during college. This program creates a cohort group among the upperclassmen mentors and freshmen and provides leadership development for all involved.

Contents

Chapter 1
Issues in Higher Education and Science, Technology, Engineering, and Mathematics (STEM)

In the science, technology, engineering, and mathematics (STEM) fields, students often struggle with content while adjusting to the college experience, and underrepresented students often encounter additional challenges (Nelson, 2007; Martin, Stefl, Cain, & Pfirman, 2020; Leary, Morewood, & Bryner, 2020). Higher education institutions strive to help students experiencing these difficulties succeed and have positive educational experiences. Researchers find that supportive advising, culturally inclusive programming, mentoring, and strategic support systems can improve retention and enhance the student experience during college (Astin, 1993; Seidman, 2012; Castellanos & Jones, 2003; Kuh, Schneider, & Association of American Colleges and Universities, 2008; Renn & Arnold, 2003; Tinto, 1993; Nerio, Webber, MacLachlan, Lopatto, & Caplan, 2019, Leary, Morewood, & Bryner, 2020).

The design of the higher education system, though, hinders some groups from persisting through college (Walker & Dixon, 2002; Martin, Stefl, Cain, & Pfirman, 2020; Fries-Britt, & White-Lewis, 2020). Astin's (1984) development theory of student involvement recommends formulating effective retention policies because "the effectiveness of any educational practice is directly related to the capacity of that policy or practice to increase involvement" (p. 298). Effective policies enable students to have positive interactions with their campus environment and give administrators and faculty opportunities to help students achieve their personal and career goals and succeed in the STEM fields.

History and Purpose of Higher Education and the Evolution of the College Student

The first higher education institution in the United States in 1636, which was Harvard, supported the goal of educating the elite. Later, higher education is defined as a more formal system to support freedom and democracy (Dewey, 1916). This

C. Harrell, D. G. Capco, *The STEM Pathway and Student Retention*,
https://doi.org/10.1007/978-3-030-65865-6_1

philosophy embraced access to higher education through the G.I. Bill of Rights, 1965 Higher Education Act, affirmative action programs, the establishment of more colleges during the 1970s, and the admission of more women during World War II (Gandara, Horn, and Orfield, 2005; Bogue & Aper, 2000). Although there was progress, the current system does not meet the need for postsecondary education" (Gandara et al, 2005; U.S. Government Publishing Office, 2020; Jungblut et al., 2018). The educational system faces the challenges of addressing the needs in STEM fields and handling the increasing numbers of students seeking access to higher education. Legal issues within higher education have attempted to address some of these challenges. The focus here is not the court cases within higher education, but seminal cases that provide a context for the current practices within higher educations.

Legal Issues in Higher Education

Legal issues arise from past practices in higher education. The Carnegie Commission in 1973 discussed controversies over practices concerning student access and academic success (Mayhew, 1973). Higher education began to be viewed as a beneficial investment and a framework for developing policies to increase access to higher education and a range of individual and societal benefits were recognized. Now educators considered who should benefit from higher education, who should pay, and provided a recommendation that one-third of the institution's cost to be covered by tuition. This information has been used to develop public policies to include the government, students, and taxpayers.

This funding distribution, however, continues to be influenced by court decisions forcing universities to evaluate their funding models and policies for students. Affirmative action and other issues related to student diversity and access to education also are impacted by court cases, such as the Supreme Court decisions in Grutter v. Bollinger (2003) and Gratz v. Bollinger (2003). These cases highlight the connection of racial and ethnic diversity in admissions to higher education. The plaintiffs in these two cases challenged the admissions policies of the University of Michigan Law School, which considered race in its admissions practices, and the Office of Undergraduate Admissions, which assigned points based on race and ethnicity. In June 2003, the Supreme Court decided that the law school's inclusion of diversity in its admissions policies was constitutional but deemed unconstitutional the Office of Undergraduate Admissions' practices that explicitly assigned points based on racial ethnicity. The court also ruled that state and federal funds for education access could not be allocated based on race or gender.

Since these and other rulings, universities have had to find different ways to expand access and funding to diversify student populations. In the *Harvard Business Review*, Bowen (1999) discuss changing trends in admissions and funding policies although race-sensitive admissions policies have existed for more than 30 years U.S. Government Publishing Office, 2020. Several reasons are discussed about why diversity should be valued, including that students of various races feel that diversity on campus and in the classroom enriches their college experience (Bowen, 1999). These and other policies influence who has access to higher education and the STEM fields.

Issues in Science Education

Concern has arisen over the United States' competitiveness in STEM fields. While the STEM definition proposed by Gonzalez, Kuenzi, the Congressional Research Service, and the Library of Congress (2012) includes science, technology, engineering, and math, although some employ broader definitions. For example, the National Science Foundation (NSF) includes psychology and the social sciences along with the core sciences and engineering (President's Council of Advisors on Science and Technology [PCAST], Washington, DC, 2012).

In contrast, others, such as the U.S. Department of Homeland Security and U.S. Immigration and Customs Enforcement, use narrower definitions, excluding the social sciences and focusing on mathematics, chemistry, physics, computer and information sciences, and engineering (PCAST, Washington, DC, 2012). While various government agencies administer STEM education programs, the following three account for four-fifths of federal STEM education activities: the NSF, the Department of Education, and the Department of Health and Human Services. This book will refer to science, technology, engineering, and math as representing the STEM field.

The PCAST report discusses areas of concerns in the STEM field, highlighting the racial and gender academic-achievement gaps in certain outcomes and challenges related to teacher quality in STEM education in kindergarten through 12th grade (K–12). Other concerns include international assessment rankings and the number of foreign students enrolled in STEM programs, of whom many return to their home countries. The report also notes issues in global STEM education, particularly the lower number of STEM degrees earned in the United States compared to other countries. The final area of concern is the lack of qualified individuals to enter the U.S. STEM labor supply.

This report discusses the need to address the achievement gap because, as some observers argue, underrepresented groups "embody a vastly underused resource and a lost opportunity for meeting our nation's technological needs" (PCAST, 2012, p. 23). Solutions include increased financial support for minority undergraduate STEM students, improvements to STEM teacher preparation, more offerings of advanced STEM courses, academic advising for minority K–12 students, increased transitional support to graduate school in STEM fields, and more minority research assistantships in STEM areas. These solutions can also help address concerns regarding women in STEM.

Due to changing trends among traditional college students, many students are unprepared to succeed in college, especially in STEM majors. Students struggle with introductory math and science courses, and STEM majors have low persistence rates (Graham, Frederick, Byars-Winston, Hunter, & Handelsman, 2013; Nerio, Webber, MacLachlan, Lopatto, & Caplan, 2019, Fries-Britt, & White-Lewis, 2020).

In 2012, President Barack Obama and the NSF, Labor Department, and American Council on Education launched a national initiative to improve STEM education and increase the number of job candidates prepared for STEM fields (Charleston, George, Jackson, Berhanu, & Amechi 2014). The program objectives included producing excellent STEM teachers and increasing the number of STEM degrees awarded.

The interactions between high school and college impacts STEM success. Although Maltese and Tai (2011) report that the number of bachelor's degrees earned has tripled over the past 40 years, this is not the case in STEM fields (Cutright, Evans, & Brantner, 2014). Accordingly, the NSF Committee on Science, Engineering, and Public Policy and the National Research Council (Charleston et al., 2014; Cutright et al., 2014) advocate for rigorous math and science curriculum in schools to prepare for the STEM workforce. These measures share the assumption that more students will work in STEM fields if they take more science and math classes. Sullins, Hernandez, and Fuller and Tashiro (1995) conclude that a significant predictor of student success is students' interest in staying in a particular STEM field (Martin, Stefl, Cain, & Pfirman, 2020) Positive experiences in high school science can motivate students to enter STEM areas in college, but once in college, students often leave the sciences due to negative experiences and a loss of interest in STEM courses. Other factors that impact success in STEM fields include finances, which affect both college persistence and time to degree completion (Ishitani, 2006; Fallon, et al., 2020). Individuals from various racial groups display attraction to STEM majors, so it might be important to increase interest in these areas among all student groups from kindergarten through college (Hilton & Lee, 1988; U.S. Government Publishing Office, 2020; Fallon, et al., 2020). Another factor influencing college students' success in STEM is the number of science classes completed in high school, which is positively associated with earning STEM degrees. Twelfth-grade students with an interest in STEM fields are three times more likely to earn STEM degrees than those planning to pursue different majors. Students who deeply desire STEM majors stay in those majors. College students who fail classes, change majors, or have children before degree completion tend to end up in non-STEM fields. However, students involved in work-study programs are no less likely to complete STEM degrees.

Framing the presentation of math and science material to make it more relevant to high school students' daily life could increase their desire to take STEM courses. STEM information should be personal, local, and relevant. For example, "rather than including chemical formulas for 'classic' reactions or discussing the effects of sea level rise on distant locals, [teachers should] focus the discussions on chemical processes in human digestion or environmental analysis of a local stream" (Maltese & Tai, 2011, p. 900). This approach to engage students by showing how science and mathematics are involved in their lives is reported to result in higher levels of interest than classes that do not emphasize these features (Maltese & Tai, 2011).

Moreover, when teachers highlight science careers, students are more interested in STEM areas. It, therefore, is concluded that a continued focus solely on achievement and enrollment in STEM majors is not sufficient to retain students in STEM fields. Instead, development, engagement, and encouragement of interest in STEM might be a better way to increase persistence in the STEM pipeline, which is the transition from high school to college and into the STEM workforce (Falloon, et al. 2020).

Scientific journals call for comprehensive changes in the educational structure to retain and graduate more students in STEM majors; U.S. Government representatives and other scientists have proposed recommendations to graduate students with STEM skills (Mervis, 2009a). It is important not only to support legislation regarding STEM education but also to implement processes to provide the finances, staff, faculty, and programs to carry out the legislation. For example, Xavier University, a historically Black college in Louisiana, has sought to improve its STEM curriculum since the 1970s and has continued this intervention into the present day. Xavier's continuous efforts have resulted in 62% of students enrolling in STEM majors and more African American students going to medical and dental school (Mervis, 2009b).

During the late 2000s, the declining economy caused many schools to cut their budgets and provide less support for STEM programs (Mervis, 2009c, 2013). Many students changed their college decisions based on the financial aid they received. Even in graduate STEM programs, the amount of financial support received by all students, including minorities, decreased. In 2010, only 51% of African Americans and 64% of Hispanics in STEM programs graduated debt free compared to 73% of Whites and Asians. Moreover, 27% of African American women—who are double minorities (women and racial group) in STEM fields—graduated with more than $30,000 in debt in contrast to only 10% of White and Asian women (Mervis, 2013).

These same issues impact graduate students in STEM education. Graduate students receive various types of support, which influence how much debt they accrue. Research assistantships often continue during the summer and provide funding for research that students use in their degrees, whereas teaching assistantships pay for efforts other than research and might not extend into the summer. These differences influence graduate students' needs for loans and lengthen the time it takes to degree completion. Mervis (2013) suggests that these disparities are affected by factors such as low socioeconomic status, family responsibilities, and lower institutional support (e.g., stipends and faculty grants). Training programs that recruit a diverse array of students may assist with some of these issues. These training programs include financial support to provide students relevant experiences in STEM-related settings. These resources assist with success in higher education.

Success is difficult in today's society without the training provided by higher education. Studies show that individuals with college degrees tend to earn more and live longer. For example, in 2008, the NSF launched a training program called the STEM Talent Expansion Program (STEP) to recruit, retain, and graduate more students into STEM fields. A notable research project by Bridgewater State University in 2010 received a $5.1-million STEP grant to implement a variety of programs in STEM pathways. The Student Retention Enhancement across Math and Science program led to simultaneous structural changes in all introductory science and mathematics courses. In addition, the university developed peer-led, cooperative-learning programs for courses, including first-semester introductory biology, calculus, and computer science classes and the first two semesters of chemistry and physics. These courses are recognized as tripping points when students often struggle or fail to enter the major. The exploration of these training programs and other factors impacting the STEM field provides a foundation for the discussion of issues in later chapters. This also suggests the importance of the STEM pipeline.

Arizona STEM Pipeline

Since our experience is from the State of Arizona our detailed understanding is for the State of Arizona in the United States. An issue in this STEM pipeline is challenges that arise from the K–12 and transition to the college level (Tawbush et al. 2020). This information is an important part of the STEM pipeline and the transition from one phase of the process into the STEM workforce. It is worthwhile to discuss issues related to underrepresented groups to better define the challenges within the STEM fields

It is important to understand the perspective of some underrepresented groups regarding STEM. The State of Black Arizona, Arizona Community Foundation, Arizona Public Service Company, & ASU Office of Public Affairs (2012) attitude survey supports five conclusions. First, being female has a positive relationship with pursuing STEM careers. Second, recent evidence indicates growing enrollment by African American women in STEM majors. Third, age has a positive relationship with pursuing STEM majors. Fourth, underrepresented groups often do not consider all vocational options equally; but when they do so, African Americans are more likely to choose STEM majors (State of Black Arizona, Arizona Community Foundation, Arizona Public Service Company, & ASU Office of Public Affairs, 2012). Fifth, when families support students' efforts to pursue STEM careers, African American students are more likely to succeed.

Addressing these issues leads to more underrepresented groups involved in the STEM field, which supports success for the STEM pipeline. Arizona's percentage is lower than the national percentages for STEM. For example Arizona's six-year graduation rate is significantly different from the national average. The state graduation rate for two-year institutions is higher than the national rate, 44% vs. 31%. The graduation rate for four-year institutions, however, is 56% nationally but 36% in Arizona. Among underrepresented student groups, it is 44% for Hispanics, 21% for African Americans, and 34% for American Indians in Arizona (F9, 2010; State of Black Arizona, Arizona Community Foundation, Arizona Public Service Company, & ASU Office of Public Affairs, 2012). On the National Assessment of Educational Progress science test, Hispanic and Black students score 50% lower than White students (National Center for Education Statistics, 2010a, b; State of Black Arizona, 2012).

It can be seen from the statistics that the percentages for success in STEM is lower in these states compared to the national percentages, which inhibits the STEM pipeline.

High school counselors should encourage students to enroll in college prerequisite courses (e.g., social sciences, chemistry). Community college programs, such as dual enrollment, Advanced Placement tests, and International Baccalaureate courses, expose students to college-level coursework, preparing them for the transition to college (Arizona Minority Education Policy Analysis Center (AMEPAC, 2007), Nerio, Webber, MacLachlan, Lopatto, & Caplan, 2019).

The gaps in STEM education can be filled by community colleges, which can serve as a vital link in the four-year college journey and equip students to transfer into four-year institutions. For example in Arizona, the Maricopa Community College District (MCCD) is one of the largest in the nation, and 10 other community college districts throughout the state allow residents to take college classes. Student success in STEM programs has led MCCD to secure NSF grants to focus on the recruitment, retention, and transfer of STEM students into four-year institutions (Hale, 2011).

Of all U.S. undergraduates taking freshman-level community college classes, about 55% are from ethnic minority groups (State of Black Arizona, Arizona Community Foundation, Arizona Public Service Company, & ASU Office of Public Affairs, 2012).

Evidence indicates that African Americans are the least represented group in STEM fields; they make up less than 10% of all students enrolled in STEM undergraduate programs and all bachelor graduates (McSherry, 2005). A possible solution for the STEM pipeline is to add more math and science classes at the K–12 level. At the college level, peer mentoring presents a potential solution.

Peer Mentoring

Gardner and the Association of Governing Boards of Universities and Colleges (2001) discuss successful strategies to address retention issues, and Kuh et al. (2008) identifies high-impact measures to improve student retention (Nerio, Webber, MacLachlan, Lopatto, & Caplan, 2019, Leary, Morewood, & Bryner, 2020).

During the first year of college, students find friends and role models, learn how to socialize, and undergo identity development and emotional changes. One recommended strategy to assist with the issues in STEM, therefore, is peer mentoring, which is shown to increase student persistence and retention. Data reveal that students involved in peer-mentoring programs have higher retention and graduation rates. Peer mentors who recently experienced many of the same changes provide new students with activities that create a sense of belonging, safety, and reassurance. Fellow students are trained to help first-year students deal with many issues they encounter (Gardner & the Association of Governing Boards of Universities and Colleges, 2001; Kuh et al., 2008; Leary, Morewood, & Bryner, 2020).

Astin (1975), Gardner and the Association of Governing Boards of Universities and Colleges (2001), and Kuh et al. (2008) also advocate for peer-mentor programs and certain mentoring theories that provide a foundation for successful programs, including formal, informal, peer, group, and self-managed mentoring. Mentoring programs are shown to aid in student retention across disciplines and countries (Beltman & Schaeben, 2012; Chester, Burton, Xenos, & Elgar, 2013; Hall & Jaugietis, 2011; Mavrinac, 2005). Even enterprises that incorporate mentoring into their business models find that they result in improved employee retention rates and a sense of belonging among mentors and mentees (Cintron, Oosahwee, & Shotton, 2007). First-semester college students face a different scenario than employees in the business world due, in part, to age-related factors and the biological, emotional, and physiological changes students' experience. Some might be unprepared for the differences between high school and first-year college. It, therefore, "is imperative to train and monitor peer mentors to ensure that expectations are clearly understood and that they acquire the appropriate tools and skills for success" (Cintron et al., 2007, p. 98).

Cintron et al. (2007) emphasize that the selection of peer mentors is critical and that a screening process should be developed to ensure the selection of qualified individuals. Research shows that peer-mentoring programs, theories, and approaches achieve success across populations, disciplines, and institutions. A comprehensive retention program considers the population and its problems and needs and offers a 360-degree experience for all involved, which can improve retention rates. A program to achieve such impacts among students is described in more detail later.

The following chapters review current issues in the K–12 system as students transition from high school to college. Common themes are socioeconomic status, under-preparedness, family support, geographic limitations, academic preparedness, and fit with school, which are addressed in Chapter 2. Chapter 3 discusses success and how to acclimate first-year students to the college experience. Chapters 4 and 5 cover the orientation of first-year students to STEM and student success. Chapter 6 emphasizes the importance of support programs that assist first-year students' transition into college STEM fields, particularly mentors' significant role in shaping the college experience. Chapters 7, 8 and 9, focus on the role of peer mentoring in STEM fields and the transition to the sophomore experience, while Chapters 10 and 11 review the course content of the mentor experience.

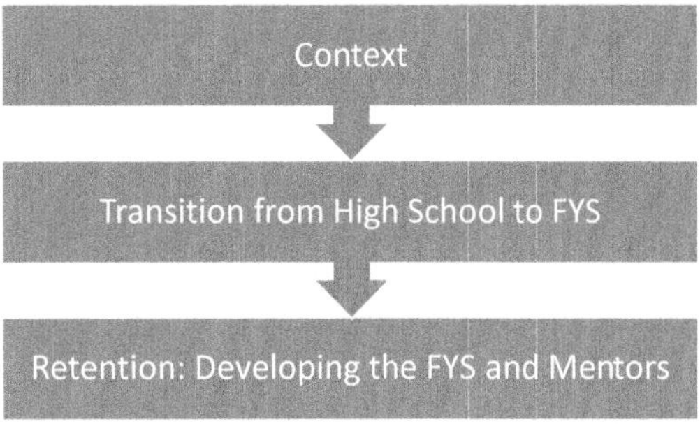

Fig. 1.1 Illustrates the three overarching areas of this book which are the context, transition from high school to the first year student, and retention

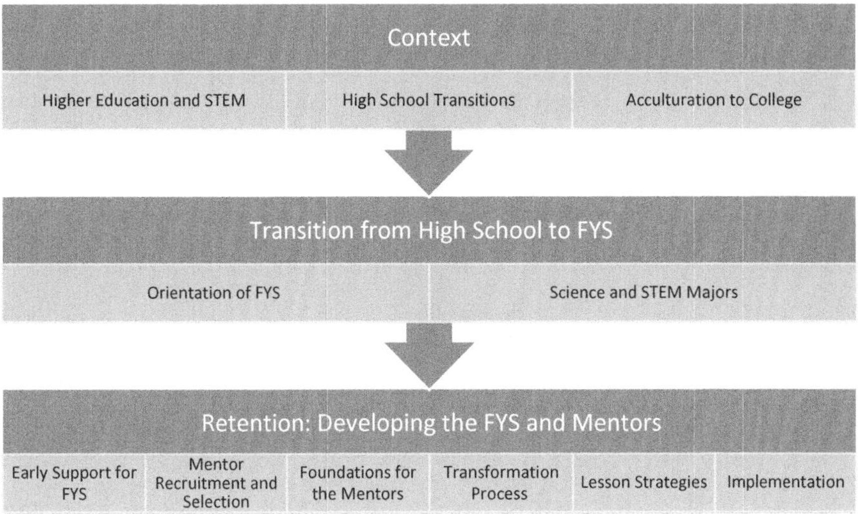

Fig. 1.2 Provides a flow of this book and within each one of these overarching areas there is a progression of phases

Chapter 2
The Transition from High School to College

The transition from high school to college provides an opportunity for peer mentors to assist first-semester students. Students transitioning to college have completed at least two cycles of school advancement, from elementary school to middle school and from middle school to high school. The messages received from family and society imply that high school graduates are ready for college, but their state of readiness depends on their college preparation during their K–12 experience (Maltese & Tai, 2011; U.S. Government Publishing Office, 2020).

Research indicates that students' gender, ethnicity, first-generation college status, and socioeconomic background, as well as the geographic location of school districts, affect students' academic preparedness (Ishitani, 2006; State of Black Arizona, Arizona Community Foundation, Arizona Public Service Company, & ASU Office of Public Affairs, 2012; Sullins et al., 1995, Martin, Stefl, Cain, & Pfirman, 2020). Some of the skills needed to successfully transition to college include time management, study skills, and coping mechanisms. Many of the necessary skills can rapidly be taught if students are willing to recognize their need for these skills. Students who feel they breezed through high school are often unwilling to recognize this need, though, until it is too late. When first overwhelmed, students fall back on their high school behaviors, which are the only ones they know, and without intervention, many of these students will not persist to the sophomore year. If their initial performance does not lead them to drop out of college, they often switch from STEM majors (Astin, 1975; Gardner & the Association of Governing Boards of Universities and Colleges, 2001; State of Black Arizona, Arizona Community Foundation, Arizona Public Service Company, & ASU Office of Public Affairs, 2012).

C. Harrell, D. G. Capco, *The STEM Pathway and Student Retention*,
https://doi.org/10.1007/978-3-030-65865-6_2

Issues in the Transition from High School to College

Preparation for college begins in kindergarten and continues through grade 12 (K–12). The K–12 system has a continuous evolution, but students often find the transition to college more challenging (Graham et al., 2013; Maltese & Tai, 2011; Martin, Stefl, Cain, & Pfirman, 2020; Tawbush et al. 2020).

Students might feel isolated and unprepared for the academic and personal challenges. They drop out of school for reasons such as finances, homesickness, and difficulties handling the academic rigor of college. Alcohol and drug use, sexual activity, and institutional fit are other factors that may hinder student success in the transition from high school to college (Fromme, Corbin, & Kruse, 2008). Students and colleges may be unaware of the importance of institutional fit, which includes the college's size, racial climate, and rigor. If students encounter unanticipated signals from the institution indicating a lack of a welcoming or inclusive environment, this may signal a lack of institutional fit (Griffith, 2010). These and many other factors influence the college transition experience.

Interventions

Venezia and Jaeger (2013) have designed several interventions to improve college readiness to assist with the college transition. These interventions range from academic preparation and understanding college financial aid to psycho-social and behavioral support to develop habits of mind, such as organizational skills, forethought, persistence, and resiliency Leary, Morewood, & Bryner, 2020).Non-academic factors in college readiness include socioeconomic impacts and college fit. Fit, again, encompasses student preparedness and what colleges have to offer, including the faculty–student ratio, cost, institution size, counseling and advising services, and diversity of the student body.

The college–home distance is especially important for underrepresented students. Some students do not move "further away than the closest community college or broad access university because they need to stay close to home to contain costs or to help their family" (Venezia & Jaeger, 2013, p. 121). These students often struggle with their goals, sense of being, and connectedness to post-secondary school institutions. Students with poor fit are less likely to graduate from college than other groups. In a study on 68 public colleges and universities and 21 flagship institutions in four states, Bowden, Abhayawansa, and Bahtsevanoglou (2015) found that poorly matched students are less likely to succeed than well-matched students. This effect may be stronger among underrepresented populations.

Underrepresented Populations

Low retention rates often affect minority students, who experience numerous issues in the college transition Fries-Britt, & White-Lewis, 2020). Zhang and Smith (2011) identify some influencing factors, including parents, friends, high school teachers and guidance counselors, college professors, academic advisors, college orientation programs, and first-year seminars. These factors can facilitate important support and skills development and help students develop an academic ethic, or placing one's "studies above leisure activities, studying on a daily basis, in a disciplined, intense, and sober fashion" (Pino & Smith, 2004, p. 829). Smith and Zhang (2008) find that students who possess an academic ethic in high school are more likely to transition smoothly to college. Zhang and Smith (2011) report the following:

> Over 15% of blacks and 8.3% of whites had a strong academic ethic in high school. 13% of blacks and 15.3% of whites had good study habits but did not have a strong incentive to learn much in high school. Another 6% of blacks and 10% of whites took challenging courses and did not avoid tough graders in high school. About the same percentage of blacks (65.5%) and whites (66.6%) did not have an academic ethic. (p. 838)

These findings suggest that strong preparation leads to a successful college transition. Similar trends among groups of students continue throughout college. More Blacks than Whites are likely to be first-generation college students and to have mothers and fathers who do not have college degrees. These students may not have family role models to assist with navigating the college climate (Zhang & Smith, 2011; Martin, Stefl, Cain, & Pfirman, 2020). This indicates the need for role models as part of the college transition, which can include mentors and the provision of other resources through the university. The success of the college transition helps retain students through graduation (Castellanos & Jones, 2003; Fries-Britt, & White-Lewis, 2020).

In the book *Majority in the Minority*, Castellanos and Jones (2003) identify the following six factors influencing minority student retention: (1) cultural and background variables, (2) socioeconomic status, (3) academic and acculturative stress, (4) family support, (5) campus climate and congruity, and (6) mentorship. These authors stress the importance of retention, defining it as "a process of creating, maintaining, and supporting ongoing strategies for meeting the personal, academic, social and financial needs of multicultural students to ensure academic success and graduation" (Castellanos & Jones, 2003, p. 19). Some high school students participate in college-preparation tracks, while others do not, which affects their successful college transition. Amid ongoing education reform, many states are changing high school graduation requirements to include college-prep courses in English, math, science, and social science.

A number of programs have been developed to help minority students transition from high school to college. Such programs introduce high school students to college life to address the academic deficits from high school. Since 1965, Upward Bound, Bridge, TRIO, Talent Search, and GEAR UP have offered summer programs before the start of the fall semester to help students transition to college life and establish an identity as a college student. This is an important factor in the transition and retention of first-year students. In addition, providing mentor opportunities at the onset of the college experience supports students' transition. TRIO and Talent Search, which originated with the 1998 Reauthorization of Higher Education Act, introduce K–12 students from high-poverty schools to the college experience. Established under the same act, GEAR UP provides student-support services, such as counseling, advising, and scholarships.

Despite these programs assisting the transition from high school to college, policy changes have decreased student retention at some institutions. For example, in the state of Washington, the elimination of affirmative action federal laws and institutional policies made some institutions seem less welcoming, so students chose to go to institutions with more inclusive atmospheres. Brown and Hirschman (2006) cite the results of the passage of Ballot Initiative 200 in Washington as an example of a challenge that likely caused stress for students. This initiative eliminated affirmative action in the state, leading to a significant decrease in minority student enrollment throughout the state. Research often focuses on a single college, but this study examines enrollment at several universities. Applicants express openness to leaving the state of Washington to have a better experience, running counter the tendency for minorities to stay close to home for family and financial reasons (Brown & Hirschman, 2006).

This review of the issues in the college transition provides the grounds for interventions. Suggested interventions include counseling, advising, college-preparation programs, and mentoring. Mentor programs can be especially beneficial for underrepresented STEM students. Career goals for STEM students may be pursuing

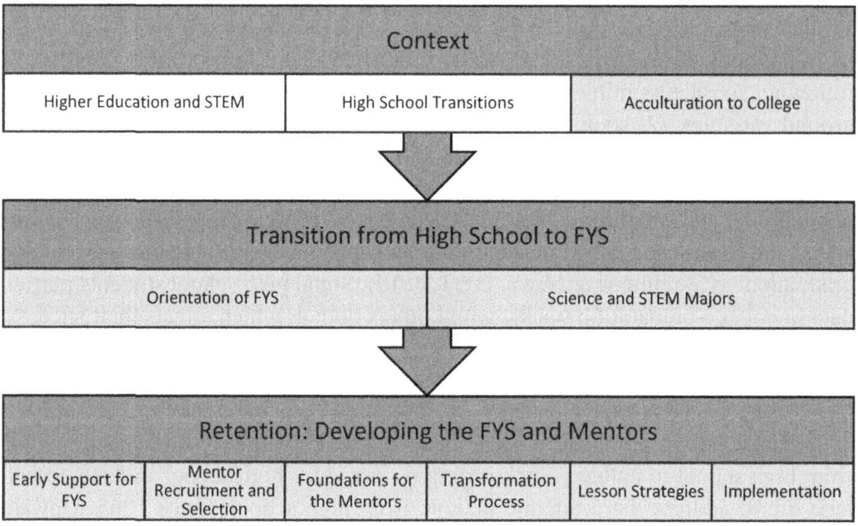

Fig. 2.1 Charts out the flow of issues in higher education and how students transition from one phase to the next

pre-health majors, graduate studies in the sciences, and entering the workforce in the science field. The details of the structure of peer-mentor programs and other interventions are explained later in Chaps. 4, 5, and 6.

Chapter 3
The Role of Identity Development in the Acculturation to College

One aspect of college students' lives that cannot be overlooked is identity development. During college, students often experience life changes and learn to understand their personal and social identities and where they fit into the university and later in society (Astin, 1975; Bowden et al., 2015; Fries-Britt, & White-Lewis, 2020). Students also develop their cultural and gender identities during the first year of college. Due to the academic performance gap, there is a need to understand racial and gender identities.

In particular, "gender is a master identity" relevant in any situation (Vantieghem, Vermeersch, & Van Houtte, 2014a, p. 363). Research associates dominance, independence, and assertiveness with masculinity and empathy, sensitivity, and kindness with femininity (Spence, Helmreich, & Stapp, 1975). Hodgetts (2008) records some differences between male and female students: female students tend to be "neat and on time, and this is a sign of a passive, compliant, malleable learning style" (p. 359). Some (Byrne, 1978) view this as a sign of an inferior learning style. There is a lack of positive female role models in textbooks and STEM fields (Skelton & Francis, 2011). In contrast, male students have an active, inquiring learning style and do not focus on teachers' expectations. Male students with high self-belief select traditional fields, including engineering, physics, economics, and mathematics, while those with low self-belief end up in non-traditional fields, such as social sciences and language.

Gender identities and racial/cultural identities are connected, and recognizing gender and cultural identities can provide approaches to address gaps in STEM fields. Consider, for example, studies on Black masculinity from the early discussion of racial identity to the complex interactions of motivations among Black men. Research provides insight into what inhibits minority students' success in college. Harper, Carini, Bridges, and Hayek (2004) and Martin and Harris (2006) discuss cultural and gender identities and identify characteristics contributing to the success of Black men, including a strong, upstanding, high moral character that values integrity, academic success, monogamy, and respect. Harris, Palmer, and Struve (2011, p. 54) describe similar masculine characteristics, such as "toughness,

C. Harrell, D. G. Capco, *The STEM Pathway and Student Retention*,
https://doi.org/10.1007/978-3-030-65865-6_3

aggressiveness, material success, restrictive emotionality and responsibility" (p. 54). These characteristics are helpful to know to best assist underrepresented students in adjusting to college and preparing for STEM fields.

It is important for students and educators to understand their own identities and who they are as individuals (Aveling, 2006; Carter & Goodwin, 1994; Mcallister & Irvine, 2000). Banks (1995) argues that individuals are not open to other ethnic groups unless they develop a positive sense of self, which includes awareness and acceptance of their own ethnic groups. White and non-White educators both need self-awareness because it helps them better understand other ethnic groups and thus be better educators (Carter & Goodwin, 1994, p. 324). Understanding racial identity theory "also assists teacher educators to identify where teachers are located along the continuum of racial identity development in order to differentiate the support for those in different statuses" (Carter & Goodwin, 1994; McAllister & Irvine, 2000; Steinert, et al., 2019). Reflecting on one's identity is a lifelong process, but doing so furthers the professional development of educators and can decrease resistance to understanding the perspectives of culturally diverse students (Han, West-Olatunji, & Thomas, 2011; Lawrence & Tatum, 1997; Rezai-Rashti & Solomon, 2008).

Helms (1990, 2008) proposes a widely accepted model of White racial identity based on the premise that "white racial identity requires progressing through two phases: a) abandonment of racism and b) defining a non-racist identity" (Han & Neuharth-Pritchett, 2010, p. 3). Helms (2008) lists the following six consecutive stages in the development of White racial identity:

1) Contact—a state of unawareness of racism and discrimination. For example, a White person might believe that everyone has equal chances for success and may have minimal experiences with people of color.
2) Disintegration—a state in which a person becomes conflicted over moral and racial dilemmas.
3) Reintegration—a state in which a person may take a step back or revisit previous stages to progress forward through the process.
4) Pseudo-independence—a state in which, for example, a White person makes a conscious intent to intellectually understand racial differences and interacts with people of color but does not internalize the understanding or the process.
5) Immersion/emersion—a state in which individuals confront their prejudices and begin to internalize their awareness of their racial identity.
6) Autonomy—a state in which White people understand racial differences and value diversity.

This information is important to assist students' college transition for several reasons. Gender, racial, and cultural awareness support the development of community and a sense of belonging, which others describe as acculturation (Ong et al., 2020; Fries-Britt, & White-Lewis, 2020). First defined by anthropologists (Redfield, Linton, & Herskovits, 1936), acculturation is often the subject in discussions on how people adjust to changing cultures when abroad. This experience is similar to when first-year students enter a college culture different from high school. Students may not be accustomed to successful higher education practices, leading to stressful experiences. The integration of students into the new culture is perceived as the most successful strategy and marginalization as the least successful approach (Warikoo & Carter, 2009). Minority students may experience similar issues, and some schools are criticized for placing too much assimilation pressure on minority students (Warikoo & Carter, 2009; Ong et al., 2020). Supportive and cooperative learning environments and bicultural development programs are reported to promote the most effective school adjustment by minority groups. For success in college, it is important to recognize students' ethnicity as part of their psychological wellbeing. A mentor program may help create an environment that facilitates identity development and provides a sense of belonging.

In mentor programs, this sense of community and belonging creates a safe place to express emotions. The mentors can use Astin's (1984) theory of involvement to create the safe place. Astin's (1984) theory of involvement holds that true involvement requires investing time and energy into academics and activities related to students' interests, thereby motivating mentors and mentees to engage in activities that aid in connecting the dots and helping first-year students progress through the transition processes. The most productive strategies for assisting first-year students are primarily developmental and sometimes transactional, which follows the philosophy of theory to practice. First-year students' basic needs should be addressed first to create a foundational sense of belonging. Mentors nurture this sense of belonging through their interactions with cohort groups. Mentors prepare to develop these groups through training in mentorship skills, the mentor's role, and leadership vocabulary. The mentors are then provided with opportunities to lead, mentor, and develop through interactions with peers and between the mentors and first-year freshmen. These strategies improve retention among first-year and upper-class students. Both mentors and first-year students strengthen their communication, critical thinking, and interpersonal skills. In addition, students become able to articulate what it means to be successful STEM students.

Research suggests that the best way to understand oneself is to use a literary approach to describe the events and feelings in one's life (Vantieghem et al., 2014a, b). Writing about these experiences consolidates them in the writer's mind, especially as the writer attempts to make the story interesting to readers. This approach is shown to aid in the development of gender and racial identities. Reflective assignments introduce mentors and first-semester students to the concepts of racial and gender identity. Students are asked questions such as how they might change their behavior or mentoring style to engage their mentees. In a final assignment, mentors reflect on their mentoring approaches and evolution during the program. Often, they acknowledge that other people have different experiences from their own, and some connect these experiences to gender and racial identity. Students may modify their approaches as they attempt to adapt to others while still developing their own gender and identity beliefs. It is the intent that this experience will start the mentors on a journey of gaining a greater understanding of themselves, which will ultimately assist their future career paths and adjustment to society. The hope is that as mentors write reflection papers and complete the final project, their internal review of the information will enable them to apply these lessons in leadership and mentoring in their interactions with their first-year students. It is important for mentors, educators, and first-semester students to gain awareness of their identities to create the sense of belonging necessary for a successful college transition.

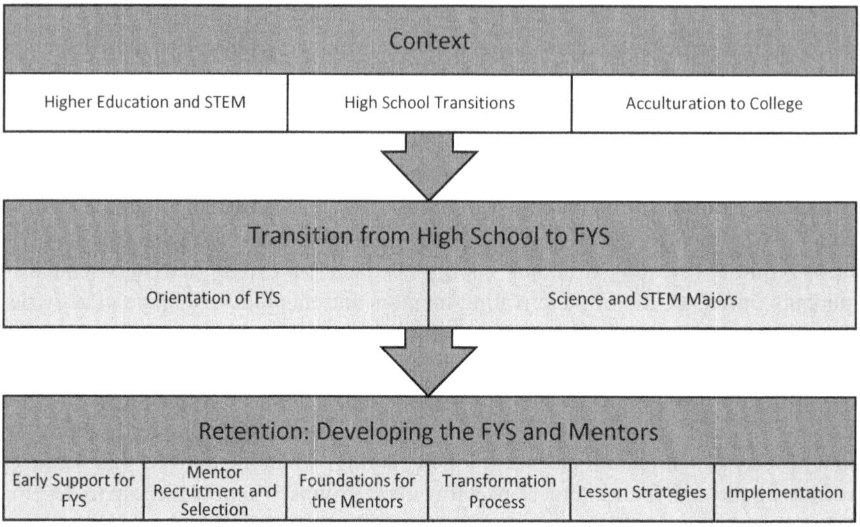

Fig. 3.1 Indicates the context and the progression through the acculuraltion to college

Chapter 4
The Orientation of First-Year Students to Success: Recognizing the Need for Change

In *Declining by Degrees* (Glasser, Merrow, & PBS Home Video, 2005), we see students at the beginning of college. The film shows images of students moving into residence halls, buying college T-shirts, and purchasing book-bags and books. They are excited to live the American dream of a college education and embark upon a new life. The students discuss their interactions with faculty, work–life balance, mentoring, and success. What is first-year student success?

There are many different views of success. This context is important for first year students to understand "the need" for the need to change. It may be a great experience, high school preparation, advising, finances, psycho-social fit, academic and social support, global relation, and mentor programs. Due to the class training and preparation in high school, the K–12 experience of first-year students is a predictor of success but not the potential for success, (U.S. Government Publishing Office, 2020). One predictor of success is advising due to the teaching and learning that takes place within this relationship. Another predictor of success is finances. As discussed in Chap. 1, reduced financial aid, rising tuition costs, and the need to balance work with class and studying can impede student success. Psycho-social fit, as discussed in Chaps. 2 and 3, strongly affects student success as do academic and social support, which include mentoring programs.

Global relations is an important predictor of success. Global relations may include interactions with international students and those from other areas as well as exposure to global issues not introduced in high school. This factor is important due to today's global society and the likelihood that students will change careers several times during their lives, including switching academic majors and changing jobs after graduation. If institutions address these factors, the potential for success increases. Democracy 2020 (Kotamraju & Blackman, 2011) describes these same guidelines for colleges and universities to prepare students for success in the workforce.

The literature and the film *Declining by Degrees* (Glasser et al., 2005) provide suggestions to support student success. Faculty involvement is additional predictors of success. Among the recommendations for faculty, instructors should encourage

C. Harrell, D. G. Capco, *The STEM Pathway and Student Retention*,
https://doi.org/10.1007/978-3-030-65865-6_4

students to talk with them in small groups so that they can realize that faculty are approachable, are available for career direction, and can provide other perspectives on life. These conversations also provide opportunities for students to develop relationships with faculty who might mentor them and write reference letters for them. The faculty interactions make science more real rather than a subject studied in the classroom or in the field. These interactions historically have occurred during faculty office hours but can also take place after class, during residential hall programming, in conference rooms, or virtual platforms such as Zoom or Google hangouts Martin, Stefl, Cain, & Pfirman, 2020; Fries-Britt, & White-Lewis, 2020).

Like a good movie, these interactions make science come to life. The narrative of scientific progress and the complexities of scientific experiences and career choices are unraveled and demystified. These academic interactions and explorations show students what college and success are all about. Social events for first-year student faculty can provide another channel for the provision of academic and social support that promotes student success. Such social events for students may include honors faculty recognition, student poster sessions, family weekends, and homecoming.

At the beginning of this chapter, the context was provided for student success. Students, colleges, and universities may define success differently. Students may define success as meeting people with similar interests, passing their classes, or, during the first week of the first semester, even remembering where they parked on a large campus. In contrast, colleges and universities measure success based on persistence and graduation rates. Higher education institutions are evaluated and ranked, notably by *U.S News and World Report*, based on student enrollment, employability, persistence, and retention. Given the changing educational trends nationally, it is in institutions' best interests to teach new students what it means to be successful college students, including their personal and professional goals. Institutional programs may also aid the transition process by addressing perceptions students have acquired from home, media, and friends about what it means to be a college student and to be successful.

Gardner and the Association of Governing Boards of Universities and Colleges' (2001) research is the first to point to the importance of programs for the success of admitted first-year students. His article "Focusing on the First-Year Student" serves as a reminder that adjustment to college life influences students' academic success and contributes to institutions' success in meeting their fiscal and mission goals (Gardner & the Association of Governing Boards of Universities and Colleges, 2001). Gardner and the Association of Governing Boards of Universities and Colleges (2001) also discuss the importance of programs to support the transition from the first year to the second year of college and explains that strengthening relationships between colleges and high schools can improve student academic success in college. Gardner and the Association of Governing Boards of Universities and Colleges (2001) identify factors that contribute to student success, including academic preparation, commitment, involvement, interaction, and active learning. First-year students also benefit from orientation, seminars, learning communities, service learning, academic support services, health education, and the recruitment of upper-class students for mentoring (Gardner & the Association of Governing Boards of Universities and Colleges, 2001).

First-year student programs, including summer and pre-semester offerings, should also address the specific needs of underrepresented groups. Financial, staff, and academic support should be provided throughout the first year of college to assist with the successful transition of underrepresented groups (Seidman et al. 2012; Gardner & the Association of Governing Boards of Universities and Colleges, 2001; Slovacek et al., 2011; Fries-Britt, & White-Lewis, 2020). Another resource that supports student success is mentorship programs, which research shows foster loyalty and a sense of belonging and support student retention. Key factors to the success of these programs are commitment, genuine care, mentees' perceptions, and relatable mentors (Cintron, Oosahwee, & Shotton, 2007). Information about student success can be disseminated through comprehensive mentorship programs that teach first-year students about college student success. This is especially important in the STEM disciplines because these curricula often leave little room for exploration compared to other disciplines.

In addition to benefiting from mentor programs, students are more likely to be successful when they recognize *"the need" for the need to change*. Doing so enables first-year students to adjust to college life. Without this realization, they might continue to practice the same habits as in high school, which are no longer sufficient due to the increased academic workload, outside jobs, time demands, and overall health and wellbeing in college (Gardner & the Association of Governing Boards of Universities and Colleges, 2001). Most first-year college students, therefore, need to recognize that they need to learn new skills and change their habits to be successful in college. Often, they become aware of this need to change after performing poorly on their first exams, experiencing illness or homesickness, or feeling overwhelmed by all the adjustments needed for the first year. At this point, first-year college students are more receptive to institutional programs, including mentoring, academic support, health and wellness opportunities, and other resources available on campus. All the previously listed constituents in this chapter view success as performing well during the first year of college. If these factors are considered, then students will be on the path to success. This progression is illustrated in Fig. 4.1.

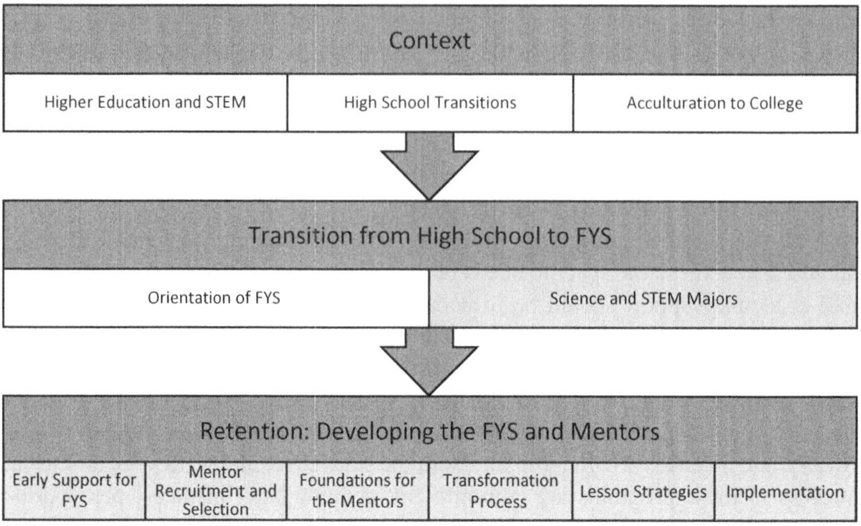

Fig. 4.1 Illustration showing the transition from acculturation to the orientaiton of first year students (FYS)

Chapter 5
The Orientation of First-Year Students to Science and STEM Majors

Preparation for STEM majors requires pre-college-level enhancements. Through interviews, Tofel-Grehl and Callahan (2017) examine the elements of the STEM pipeline at the high school level and find that teachers have specific expectations of students who excel in STEM areas. Several resources can aid the college transition for first-year STEM students. Financial resources, staff, academic support, and programming contribute to a successful transition during the first semester (Seidman et al. 2012; Gardner et al., 2001; Slovacek et al., 2011). Initial post-secondary-school support includes advising that encourages students to enter STEM fields of studies and to declare STEM majors early in their college careers to minimize the time and money required for degree completion (Frehill, 1997; Chen and Upah 2020).

Perceptions and Issues in K–12 Science

Earlier chapters cover major concepts in K–12 education, and it is important to review their connection with STEM. Wang (2017) identifies the following three factors important for K–12 students' success in STEM majors in college: intent to major in STEM fields, high school math achievement, and initial secondary experiences, including academic interactions and financial aid. These findings, along with other studies, support that twelfth-grade math achievement and exposure to math and science courses have the greatest influence on intent to major in STEM in college (Wang, 2017; Wolniak, 2016; Ramsay-Jordan, 2020). According to the National Science Board (2010), the employment rate in science and engineering increased by 3.3% annually from 2004 to 2008, while the employment rate in all occupations rose by only 1.3% annually during that same timeframe. Minority, female, and low-economic status students have less representation in STEM fields than majority groups (Herrera & Hurtado, 2011; National Science Board, 2010). For both majority and minority groups, the post-secondary-school experiences that inhibit success in STEM majors include low financial aid, family and job responsibilities, graduate

C. Harrell, D. G. Capco, *The STEM Pathway and Student Retention*,
https://doi.org/10.1007/978-3-030-65865-6_5

degree expectations, and interventions that demand taking remedial courses in math, reading, and writing (Wang, 2017; Wolniak, 2016; Ramsay-Jordan, 2020).

Wang (2017) expands this to include math and science courses as strong predictors of success in STEM majors. Minority groups are likely to have less exposure to high school math and science courses, which can discourage the intent to pursue STEM degrees. As one mechanism to address the low rate of minority students choosing STEM degrees, Wang (2017) suggests that early exposure to these majors has a greater effect on minority groups than other groups because it improves math self-efficacy beliefs, thereby increasing intent to major in STEM fields (Bandura, 1997; Wang, 2017). This exposure should occur early in K–12 education.

Expectations of First-Year STEM Students and Faculty

Preparation for STEM majors requires pre-college-level enhancements. Through interviews, Tofel-Grehl and Callahan (2017) examine the elements of the STEM pipeline at the high school level and find that teachers have specific expectations of students who excel in STEM areas. Teachers at STEM schools believe that their students have mindsets intellectually and socially similar to themselves. The teachers also believe that gifted students have an obligation to learn independently outside class-time and that doing so allows them to learn authentically free of teachers' influence and bias. Another belief is that high-performing students need major challenges and a substantial workload. The teachers highly value inquiry-based learning. In contrast, the students feel that they often struggle and have little support from teachers. Teachers' perceptions of students' ability influences student success (Borda et al. 2020; Park et al., 2020; Riegle-Crumb et al. 2020). These same student and teacher expectations influence the pedagogy of poverty that lower-income and other minority students do not receive the same classroom opportunities as their majority peers (Boaler, 2002; Haberman, 1991; Tofel-Grehl & Callahan, 2017; Tawbush et al. 2020). Often, teachers spend more time on instruction-based activities than inquiry-based learning (Haberman, 1991, Borda et al. 2020). Similar to the recommended practices of exposing K–12 students to STEM, it is also important to expose K–12 students to college, which improves their college transition and success. To expose K–12 students to STEM concepts, K–12 teachers and college faculty need training in science and math concepts and applications, and where possible, students' and teachers' enthusiasm for STEM concepts should be developed (Borda et al. 2020). These interactions promote STEM career pathways and social reliance on STEM; foster interactions between K–12 students, the research community, and local STEM professionals; and empower faculty to give K–12 students STEM learning experiences. Faculty also need continual leadership and professional development opportunities (Pinnell et al., 2013; Steinert et al., 2019).

Fuentes, Alvarado, Berdan, and DeAngelo (2014) find that first-year college students' exposure to faculty has great impacts on them by their senior year of college. This study proposes that the socialization of students and faculty enables students to acquire the knowledge, skills, and disposition needed by effective members of the college scientific community. Faculty act as mentors, passing on experience and knowledge to the next generation (Girves, Zepeda, & Gwathmey, 2005). Faculty and first-year students can interact formally in class and informally outside class through course-related interactions, intellectual discussions, holistic development, research-related interactions, advisement, and constructive feedback (Borda et al. 2020). Faculty expect that it is important for them to offer interactions and mentoring to first-year students to increase retention. It is recommended that institutions support faculty–first-year-student interactions by allocating resources to increase their contact and communication. Comprehensive programs are more likely to promote these than isolated programs.

One source of students' expectation is the feeling of entitlement, which arises from the view that students are customers in higher education. Students thus believe that they are entitled to goods and services provided by the institution. The millennial generation born from 1982 to 2002 has highlighted these issues. The early work by Morrow (1994a, b), who first uses the term "millennial," asks the question:

> Do we want to award degrees to everyone that feels entitled to this distinction and thus render the award meaningless, or do we want to reorientate students so that they realize that true skill and effort are required for achievement? (p. 344)

Numerous authors report that millennials expect to have choices, control, and instant gratification (Carlson, 2005; Rickes, 2009; Shaw & Fairhurst, 2008; Stewart & Lozano, 2009). This study conducted at a midsize Canadian university provides a comparison with U.S. institutions (Singleton-Jackson, Jackson, & Reinhardt, 2010). According to the study's recommendations, institutions should explore the balance between faculty and students in the creation of course outlines and syllabi by listing the responsibilities of both parties. Marketing teams can involve faculty in the sales pitch used when recruiting students and setting policies. A committee should be formed to create a mission statement that sets out both the academic goals and service expectations for students. Future research should investigate how faculty, institutions, and society perpetuate student entitlement and how it can be used to have a positive impact on learning (Chitiga et al. 2019).

Institutions should also consider the diverse needs within the student population. For example, students who have not yet decided their majors may not receive socialization within academic departments, so they should be provided a space for socialization. Second, for students who identify with particular races or genders, institutions could provide mentoring opportunities and access to supportive resources. Faculty may also need resources to adapt their mentoring to ensure that they are inclusive of multiple student populations (Steinert et al., 2019). Engagement between faculty and students is important to help students learn skills relevant in today's society. A study on engagement among engineering students and faculty tests this hypothesis. Faculty define student engagement by students' desire to learn on their own, involvement in discussions in and outside the classroom, and the asking of questions related to homework. Engagement includes faculty–student interactions rather than traditional lectures. Students expect faculty to engage them through their teaching style and affect. Students' engagement increases as they have an active role in labs, simulations, discussions, and projects.

Both faculty and students view engagement as having multiple dimensions and being capable of supporting various perspectives. Students view engagement as supplied by faculty, who, in turn, view engagement as an outcome of student actions. In a sense, faculty focus on learning outcomes and students on the input provided by faculty; both groups thus have different expectations. Jansen, Andre, and Suhre (2013) suggest that students' expectations may influence their success and predictors for certain outcomes.

Information about the engagement between faculty and students may also provide suggestions to enhance policy and curriculum development for first-year students. Jansen et al. (2013) provide a model of a successful college transition with the following four stages: preparation, encounter, adjustment, and stabilization. First, the preparation stage occurs in high school and through other life experiences as students develop expectations of what college will be like. Second, in the encounter stage, students arrive at college. Depending on their preparation stage, they now develop ways to adjust during the first semester of college, which is called the adjustment stage. Fourth, the stabilization stage marks the end of the college transition and should lead to student retention.

All of these transition points may create stress, which can be reduced by stress reduction practices that should be included in curriculum design. Waters, Marzano, and McNulty (2003) notes that most first-year curricula and teaching practices do not match students' preparedness. Studies on student attrition identify reasons why students leave this transition cycle, including unsatisfactory experiences, dislike for their chosen field of study, dealing with university study demands, and personal factors, such as financial problems and family issues. Students often indicate that they have difficulty understanding faculty members' expectations and the time constraints of college education (Keup & Stolezenberg, 2004; Borda et al. 2020). Around one-third of students report successfully developing effective study and time-management skills, which smooth the transition process.

Faculty expectations for student performance on exams and class assignments often do not match students' perceptions (Park et al., 2020; Chitiga et al. 2019). Course descriptions may not provide enough information for students to prepare, leading to poor performance on class assignments and exams. Students may then fall back on the study methods they used in high school. In summary, student success is dependent upon how strongly students believe in themselves before coming to college. In a cross-cultural study examining four major aspects of expectations of student readiness and perceived preparedness (Jansen et al., 2013), the four items for student readiness are information processing, collaborative learning, time management, and writing. The authors use three major equivalencies to standardize student responses. The characteristics of student expectations evaluated are students' expectations of the faculty, including that they will teach how to write essays and take notes. The second expectation is that faculty will involve or expose students to research activities. Third, students expect college to be similar to high school. Fourth, students expect that they need to employ time-management skills.

A questionnaire administered to first-year students finds cultural differences among the participants. Some differences are higher expectations for initiation and time-management skills. Other students believe they will be instructed on how to complete assignments and gain research experience in their degree programs. Higher education institutions are responsible for managing first-year students' transition to college by designing a learning environment that encourages these students and offers an infrastructure that provides their needed support (Iverson et al., 2020; Chitiga et al. 2019). This is part of bridging the gap for first-year STEM students.

Bridging the Gap for First-Year STEM Students

Several resources can aid the college transition for first-year STEM students. Financial resources, staff, academic support, and programming contribute to a successful transition during the first semester (Seidman et al. 2012; Gardner et al. 2001; Slovacek et al., 2011). Initial post-secondary-school support includes advising that encourages students to enter STEM fields of studies and to declare STEM majors early in their college careers to minimize the time and money required for degree completion (Frehill, 1997; Chen and Upah 2020). Substantial financial support gives students more time to interact with faculty and advisors. This is often needed for STEM majors and is extremely important for underrepresented groups who face additional issues, including costs for STEM courses, such as lab fees. Tuition at large research universities can also be notably higher than at other institutions (Kienzl & Trent, 2009). Financial resources provide needed support for students to succeed in STEM fields, allowing them to spend more time on their studies than in other fields (Arum & Roksa, 2011; Brint, Cantwell, & Saxena, 2012). In addition, parental involvement may increase STEM degree completion success (Moakler & Kim, 2014).

Research on the retention of minority students has not often considered American Indian students in STEM fields. One transition pathway for American Indian students in STEM fields is to study first at a two-year tribal college and then transfer to a four-year institution. Tribal colleges aim to preserve local traditions and languages, so this pathway has become an important way to maintain the enthusiasm of American Indian students. In addition, mentoring is an indigenous pedagogy for American Indian students and can strengthen the skills of students with deficits in math and science (Ward, Coles, Rich, Knapp, & Madsen, 2014a, b, c; Chelberg, 2020). While some scholars support Tinto's (1993) emphasis on assimilation, others find that persistence and academic achievement are better addressed without assimilation and instead stress local cultural practices.

Inclusive strategies are important for institutions to integrate students into college life (Rendon, Jalomo, & Nora, 2000). In addition, Tierney and Lechuga's (2004) model shows that cultural capital helps students overcome retention barriers. Several authors report that both native and non-native faculty can be effective as mentors and that racial similarities between mentors and students are not essential for positive learning outcomes (Gaddis, 2012). However, non-native faculty may have greater challenges teaching indigenous knowledge (Abrams, Taylor, & Guo, 2014). Moreover, findings show that mentor research opportunities have positive impacts on tribal college students and serve as meaningful mentoring strategies for native and non-native faculty (Chelberg, 2020). Non-native faculty develop relationships that include both trust and collaboration. Both native and non-native faculty are effective at mentoring students as long as they recognize the importance of culture and see students as individuals. A cultural value within the tribal community is caring relationships in which older adults provide students with guidance and assistance. This can also be modeled in mentoring relationships on college campuses.

Guidelines for the Future

The previous chapters discuss the important roles of the development of community among first-year students and mentors as well as students' perceptions and issues based on their K–12 STEM experiences. The literature identifies a sense of belonging (Strayhorn, 2012) as a key to educational success for all students and discusses the best combinations or best practices to create an environment conducive to success. A community or sense of belonging provides the foundation for STEM learning and can be crucial if students become disenchanted with their majors. Such a feeling typically arises during the fifth or sixth week of a traditional college semester, when students begin to question if their STEM field is the best career option. Even when students question their sense of direction, a sense of community or belonging can provide them with the security necessary to remain in college and in STEM-related career paths.

Even many honors students prepared for STEM majors begin to question whether STEM is the best fit for them. Certainly, students from ethnic-minority and low socioeconomic backgrounds commonly have deficits in preparation for STEM courses and have difficulties meeting challenges as they begin to question their fit with STEM or their institutions. Broadening STEM participation among underrepresented students can be challenging given what many faculty think is the appropriate speed of learning.

Often, many faculty who are not from low socioeconomic or ethnic-minority groups do not have a clear understanding of the potential of these groups and their developmental needs to bring them fully into the STEM curriculum. There is a belief that accelerated courses and additional assignments may solve this difficulty. However, students often state that they feel overwhelmed and defeated with the additional tasks intended to help them succeed in STEM (Park et al., 2020). Here, a sense of belonging provides security to persevere in the midst of what students view as major challenges. Here, too, a mentor program can provide effective support. Figure 5.1 shows the transition point to the next overarching area of retention.

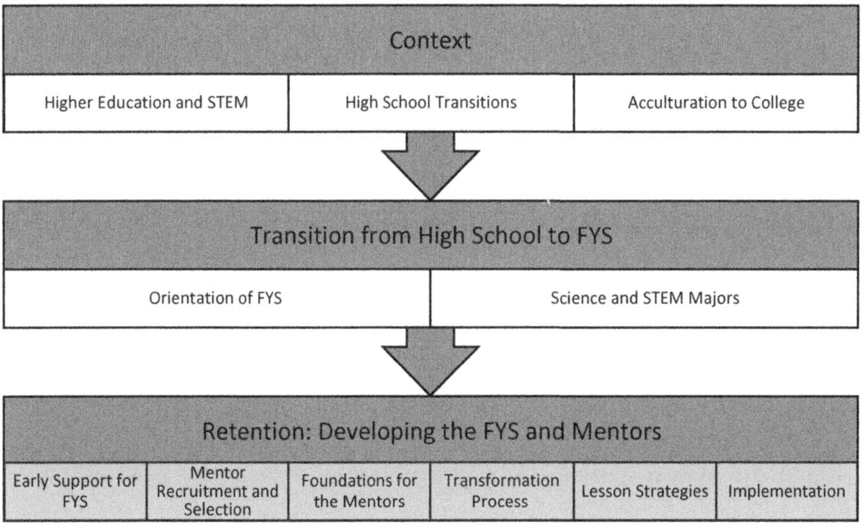

Fig. 5.1 Shows the progression from first year students to successful STEM majors

Chapter 6
Early Support for First Semester Freshmen: Peer Mentoring

Cohorts and Peer Mentors (Setting the Stage for First-Semester Freshmen)

Early support programs include cohorts, peer mentoring, and other factors to improve the student experience and student retention. Some of these factors are development of cohorts, mentoring approaches, leadership, and the peer mentor process. The combination of these areas set the stage for the first-semester freshmen and will be discussed within this chapter. The first early support factor is the development of cohorts.

Cohorts

What is a cohort, and why is it relevant? Scholars state that cohorts can be discipline- or topic-focused, and for our purposes, a cohort is a group of students within the same discipline who have a common experience (Christe, 2013; Smith et al., 2008; Gunn et al., 2020). This common experience gives them a connection when transitioning into college, exposes them to the discipline, and provides an opportunity for a group setting that creates support and encouragement (Christe, 2013; Pike & Robbins, 2019). Our program has three types of cohorts. One is a cohort of first-semester freshmen, who take several classes as a group and become familiar with the other members. The second cohort includes the first-semester freshmen and one or two peer mentors who are more senior than the freshmen. The third group consists of the mentors themselves and a more senior emeritus mentor.

Cohorts that include living-learning programs have been shown to highlight the development of self-efficacy, or a belief in oneself, and expectations for a good career outlook and work–life balance (Seidman, 2012; Astin, 1984; Bandura, 1997). These qualities available through cohort support increase persistence and retention

C. Harrell, D. G. Capco, *The STEM Pathway and Student Retention*,
https://doi.org/10.1007/978-3-030-65865-6_6

among college students (Seidman, 2012; Pike & Robbins, 2019). A study by Whalen and Shelley (2010) demonstrated these outcomes and was conducted in the fall of 2000 at a Midwestern research university. The study's goal was to examine the background characteristics, ability measures, financial support systems, and academic support mechanisms for the retention of STEM and non-STEM majors and underrepresented students in STEM majors. The dominant predictor of six-year retention and graduation was found to be cumulative GPA in the last year of high school (Whalen & Shelley, 2010). Another study, by Kalevitch, Maureer, Badger, Holdan, and Sirinterlikci (2015), in an NSF-funded program showed significant retention of both STEM and non-STEM students among students who received significant financial support over four years of college and were grouped into cohorts. In addition, there was a living-and-learning community for the students who chose to live on campus. Student retention from semester to semester was 81% or greater. Living-learning communities include the benefits of cohorts. These factors improve student retention and lead to the integration of mentoring within the cohort system.

Mentoring

There are several different types or definitions of mentoring. Some of the most common categories are formal, informal, peer, group, and self-managed mentoring (Zevallos & Washburn, 2014; Tenenbaum, Anderson, Jett, & Yourick, 2014; Spaulding et al., 2020; Chelberg, 2020). Mentoring is a learning partnership between relatively more- and less-experienced individuals (Kricorian et al., 2020). It is a process involving emotional (friendship, acceptance, and support) and instrumental (information, coaching, advocacy, and sponsorship) functions (Elliott et al., 2020). It is a relationship that becomes more impactful over time. Others discuss nurturing the social and psychological development, serving as a role model, and providing support for goal setting and future planning (Atkins et al., 2020; Kricorian et al., 2020). This definition leads to the stages of mentoring.

There are three stages of mentoring, including the beginning or introductory stage, middle stage, and transitional or concluding stage (Ensher & Murphy, 2011). This information is applicable for first-semester freshmen. Characteristics of the introductory stage include the development of trust in and the foundation for the mentor–mentee relationship. This can be achieved initially by the mentors for first-semester freshmen facilitating icebreakers and other activities for the mentees to learn more about the college experience and build community. In the middle stage, the mentors facilitate activities to build skills and encourage freshmen who seem to be struggling at this point during the first semester of college. Some activities include goal setting, time management, note taking, and mindset. In the final stage, the mentors assist the freshmen in transitioning to the second semester and sophomore year. During this stage, the freshmen are introduced to career paths, stress management, leadership opportunities, and transitional campus resources and support. This encourages the freshmen in regard to their ability to succeed in college. This experience is beneficial for all involved in the mentoring process.

The benefits of mentoring include motivation that promotes an individual sense of success and enthusiasm about the specific disciplines, which can lead to academic success and persistence (Kricorian et al., 2020). Mentors can promote interaction with an upper-class cohort of mentors and faculty (Spaulding et al., 2020). "Mentoring does not come from a guide book, set of rules, or even incentives. Mentoring comes from the heart" (Vesilind, 2001, p. 410).

Leadership Development

Why does a mentor have to be a leader? Many studies suggest that leadership is an important part of mentoring (Colvin & Ashman, 2010; Spaulding et al., 2020). In the Colvin and Ashman (2010) study, only 10 of the 96 comments referred to mentors as peer leaders, but many referred to leadership qualities and mentors' guidance of mentees. In the same study, the role of student advocate was the second highest characteristic associated with being a mentor.

Various approaches have been used to teach student leadership and mentoring. Some are one-day workshops, whereas others are full-degree-granting programs. Some train students to lead within a particular context, such as a medical students' association. Students develop leadership skills in an area where they feel committed, and in some cases, student leaders are developed in groups of students not directly interested in leadership development activities (Spaulding et al., 2020). Micari, Gould, and Lainez (2010) investigated this area of developing leadership in students not interested in development, particularly in STEM areas, based on the students' perspective. "The learning objectives for this program are pedagogy theory and research, gain knowledge of small group dynamics, gain practical facilitation skills, develop inquiry skills by engaging research on practice, reflect on practice and gain self-evaluation skills" (Micari, Gould, and Lainez 2010, p. 221). In surveys, the students reported that they felt that they had improved their cognitive skills and were equipped to interact with others in a variety of pedagogical approaches. They also gained knowledge on their own about concepts in the field because to teach it, they had to draw together the parts in their own minds.

In one study, by Merisotis and Kee (2006), on 10 years of sustained investment in collaborative leadership, a model was developed to aid in the retention of Latino STEM students. The first phase of this program began with the Model Institutions for Excellence (MIE) program. There were several phases, including investment, distribution, integration, production, and reinvestment phases. The investment phase included receiving financial and intellectual support from NSF and the National Aeronautics and Space Administration (NASA). In the second phase, the distribution phase, these investments were used to implement strategies related to the goals, including the physical infrastructure of buildings and the provision of support for STEM students through tutoring, internet access, meeting rooms, internships, career events, and research events. The changes in the physical infrastructure created an environment for student success. Technical tools, such as video conferencing and SMART boards, were also provided. In addition, during this phase, faculty created new STEM courses, and faculty in tribal colleges expanded their STEM curriculums using some of the same processes.

The third phase, integration, analyzed student outcomes, particularly changes in STEM enrollment and graduation and the number of students continuing into graduate programs. The production phase included using the practical lessons learned from the Model Institutions for Excellence (MIE) program to develop STEM programs and scientific discovery at other institutions. The final phase, reinvestment, took the intellectual capital acquired to expose other institutions to the learning strategies and lessons from this 10-year program (Merisotis & Kee, 2006).

Another program promoting leadership development was the NYC College of Technology. It developed a three-tier program of peer-led team learning leadership training, research scholars, and graduate studies preparation. In the first tier, the students were mentored by faculty and learning specialists to be workshop facilitators and gain an understanding of leadership and communication skills in their content area. The second tier involved students in undergraduate STEM research experiences and was intended to motivate them to pursue advanced STEM degrees. The third tier provided support to move underrepresented minorities into the graduate school pipeline. This study reported that the leadership and support measures were successful at guiding students to graduate school and that peer leaders experienced strong motivation to effect changes in their attitudes toward STEM. Students gained new critical thinking skills rather than only content knowledge, aiding in the development of new attitudes and helping them make informed choices about their career direction. It was suggested that these accessible pathways also helped create a STEM pipeline for minority students (Ghosh-Dastidar & Liou-Mark, 2014; Bhatt et al., 2020). Leadership and cohorts converge within the peer mentor program to create a better support system to retain first-semester freshmen.

The Role of Peer Mentors

At the many institutions that have employed peer mentors, several factors help peer mentors be effective. Mentors' ability to make impacts on first-semester freshman can be influenced by the faculty running the courses. Even when a course design has specific roles for mentors, it requires changes in the faculty to empower the mentors if the faculty come from traditional backgrounds in academia, particularly STEM fields (Jansen et al., 2013; Spaulding et al., 2020). Many STEM faculty might view mentors as teaching assistants, who might attend a class but have no real role during class time. This does not empower mentors and takes authority away from them. Mentors need to be an integral part of the classes. To facilitate the training of new mentors, former mentors from a program may assist in the training process. In the first-year-retention STEM mentoring program that we developed, returning upper-class mentors are called "emeriti mentors" and bring a greater level of enthusiasm to the mentoring process. Their enthusiasm and leadership skills provide additional support to new mentors and freshmen during the first semester. The emeriti and new mentors create the third cohort group, discussed previously in this chapter.

What role do the mentors play in this in program? As detailed in other sections, the first formal class period is largely facilitated by the mentors as they meet with the small groups of freshmen assigned to be their mentees. This initial investment of time is essential both to begin the formation of bonds between the mentor and mentees and to show that this will not be a traditional, structured class. The mentors also lead two more classes and facilitate activities in additional classes. The mentor–mentee interaction is what makes the content information in this course believable to the first-semester freshmen. This interaction assists in developing a sense of belonging, helping first-semester freshmen connect to the STEM major. Mentors facilitate study groups, assist students in developing study and time-management skills, and simply listen to problems the students want to discuss. It is the mentor who, after listening, can often say, "Yes, I made a mistake just like that" or "When that happened to me, this is how I solved that problem."

Faculty cannot provide support in the same way as mentors. Even if the faculty have the time to simply listen to the students' problems and to hold multiple study groups for them, the faculty will not have had similar experiences only one or two years prior. In fact, with the way college education has changed, faculty may never have had the same experiences as the first-semester freshmen. This is especially true because the faculty who have the time to invest in such classes typically are post-tenured faculty and significantly older than the first-semester freshmen.

Other ways to lead during the course include discussions on specific topics, perhaps doing a reading or watching a video freshman have been involved with outside class. For faculty to flip a class or to create other types of active learning requires a great deal of reorganization from a traditional lecture format. Flipped classes and active learning include student-centered activities and instruction (Michael, 2006; Michael & Modell, 2003; Wang, 2020). The addition of peer mentors to any class is much easier as long as the faculty are willing to give up some of the control they have over the traditional class period. Moreover, as noted, trying to transition freshmen from high school to college by putting them into an active-learning classroom is a big jump for some. Engaging these first-semester freshmen with peer mentors in the class provides support for this change, so by the second semester of the freshman year or by the sophomore year, they can easily continue with active-learning classes. In addition, mentors assist freshmen in adapting to different learning styles.

At any institution, there likely is a diverse array of students from many socioeconomic backgrounds. College preparation may include adjustment to active learning and other learning approaches. These students' preparation for entry into college differs and may influence how they perform in college. Particularly in the first semester of the freshmen year or even the sophomore year, the students may still be catching up from their K–12 training (Gunn et al., 2020). Just as there are many different students, there are many different learning styles and many ways to provide instruction. During study groups, mentors can assist students in being successful with different teaching and learning styles and approaches. At many higher education institutions, students often have only traditional courses. An active-learning curriculum leads to better retention of information and the ability to use information (Michael, 2006; Michael & Modell, 2003). However, having either of these as an "all or none" approach does not really serve the entire class of freshmen, especially when their high schools provide largely traditional curricula. To take students whose only instructional experience has been in traditional classes and thrust them into active-learning or flipped classrooms could easily be a disservice to them. This is especially true for first-semester freshmen who have had no role models, perhaps also have poor time-management and study skills, and consequently cannot find direction until late in the semester. This situation does not lead to retention. The mentors are there to support the freshmen to have a successful semester.

Summary on Early Support Programs: Cohorts, Leadership, and Setting the Stage

Early support programs include cohorts, mentoring, and leadership which set the stage for first-semester freshmen to be successful. The benefits of cohorts include the creation of a sense of belonging, a welcoming STEM environment, and adverse effects on characteristics related to dropping out of STEM majors. Mentoring reinforces all of these factors and happens at different levels and stages, allowing the relationships to become impactful throughout the semester. Mentoring reinforces the sense of belonging created by the cohort structure. Effective mentoring programs require that mentors understand and employ aspects of leadership. Mentors must be trained to develop their leadership skills. Focusing on leadership development may also increase the success of underrepresented groups in STEM. Based on this information, it is important to understand the role of mentors in supporting the success of freshmen during the first semester in STEM.

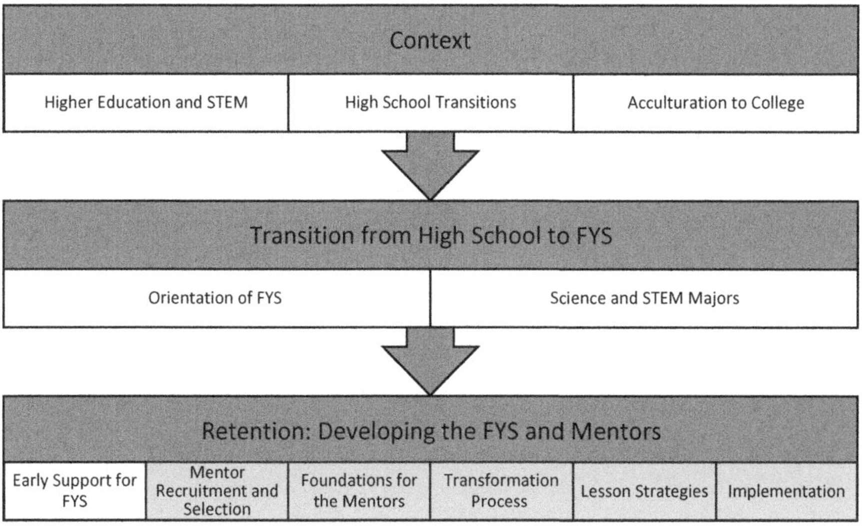

Fig. 6.1 Starts the process of retention with early support programs as the first element

Chapter 7
Recruiting Mentors and the Selection Process

There are many approaches to developing a mentoring program. Often the phases include nomination, evaluation, acceptance, training, and surveys (Alonso, Castaño, Calles & Sánchez-Herrero, 2010; Beltman & Schaeben, 2012; Black, Terry, & Buhler, 2016; Hall & Jaugietis, 2011; Holland, Major, & Orvis, 2012; Tennenbaum et al., 2014; Zevallos & Washburn, 2014; Spaulding, et al., 2020). Our program uses similar phases, and in this chapter, we describe the process while providing the rationale for each step. This should enable others to understand the process better and provide the foundation for any desired modifications. Within the training phase, engagement lays a foundation of enthusiasm for the training, which results in buy-in from the new mentors and impacts the first-semester freshmen's interest.

Start of the Recruitment Process with Nominations of Students as Mentors

The first step in acquiring a cohort of new mentors is to present the program to potential future mentors in an exciting, informative way. The importance of being a mentor and the selectiveness of the process should be clear. To be a mentor, students must be nominated, and those nominated need to fill out an application. The application form should not be too onerous and should seek information relevant to the selection process so that it does not need to be collected twice. This process should also allow the students to recognize that acceptance is into a competitive program taking only a small number of nominees from a much larger group. In addition, the process allows faculty members to write letters of recommendation for mentors, indicating that they were selected as, for example, only 50 mentors from a group of 800 upperclassmen (2.5% of the total group). Such statements used in letters of recommendation and by the students during job interviews and on their résumés provides evidence that they were in an elite group and certainly makes them stand

C. Harrell, D. G. Capco, *The STEM Pathway and Student Retention*,
https://doi.org/10.1007/978-3-030-65865-6_7

out. In such letters, followed by a brief description of the mentor program, faculty can make unique statements about the students.

Future mentors can be nominated from several sources, and likely all sources should be used to obtain enough qualified new mentors each semester. Each source has advantages and disadvantages. One source is the STEM advising office in the major. The advising office has the advantage of seeing many students in the major (in our case, more than 3000 students per year) and has access to students' GPAs, so it can ask students meeting the criteria for admission if they would like to be nominated for the mentor program. The advising office, though, gives out much advice, making nominations simply one piece in a flood of information, so it has the disadvantage of not appearing to be special or elite. If the advising office sends a bulk email to a list server, the nomination may not appear special or elite. This could result in subdued interest in recruitment among students. Starting with enthused students gives the program a better foundation.

Another source of mentor nominees can be the previous year's mentors themselves. Toward the end of the fall semester, after the mentors have experienced much of the program, they are asked to nominate two types of students: their classmates who they think are interested in becoming mentors in the next fall semester and first-semester freshmen mentees who have responded positively and enthusiastically to the freshman retention process. Some mentees nominated by the mentors might have changed their major by the next fall semester, but this is sorted out in the evaluation process, which determines if the new major is aligned with the program. For example, if a student has changed from biology to biochemistry or sustainability, these are still applicable majors.

A possible concern is that in the next fall semester, the new mentor will be a sophomore and might not be sufficiently prepared. However, there are advantages to recruiting sophomores as new mentors. The selection criteria used ensure that capable students are recruited. Moreover, the former mentees' enthusiasm for the program can create the difference in making their interaction with next year's first-semester freshmen productive. Furthermore, these new mentors recently went through and survived their first year, so their challenges and solutions are immediately applicable to the new first-semester freshmen. Just as importantly, recruiting first-semester sophomores to be new mentors positions them to be emeriti mentors in their junior and senior years if their course schedules permit. Several students have followed this path. The emeriti mentors come in with a different level of enthusiasm and wisdom. Students who have been emeriti mentors for two years often have many insights and can bring new mentors along more rapidly. If we allow only the most senior students be emeriti mentors in the fall semester of their senior year, they will graduate and depart before the next fall. Assigning sophomores to be new mentors creates a mixing pot of junior and senior emeriti mentors, which gives the process more energy.

A third source of nominations are the faculty. The advantage of this source is that students are selected based on success in the classroom and thus more serious students apply. The disadvantage is that many faculty are so engrossed in their research and other faculty duties that they often do not get around to making nominations. If the faculty make nominations based on the students they have in class, they typically wait until near the end of the semester to make nominations, risking that nominations get lost in the end-of-the-semester crunch of finals and other duties. Moreover, faculty tend to think of nominating only one or two students, not a large number of students, but what is needed are nominations of large groups of students. Students' schedules make it hard to fit into the mentor program, so, for example, out of 30 nominations, only five students might be able to fit the program into their schedule in the next fall semester.

Students need to know that faculty have nominated them. If a faculty member simply passes along a student's name to the advising office, which sends a letter informing the student of the nomination, the nomination does not have the same effect. It is more effective if the faculty member informs the student of the nomination before giving the student's name to the advising office. The following steps carried out by the faculty may lead to the greatest response from students:

(a) While the course grades and names provided by the teaching platform (e.g., Blackboard and Canvas) are used as an excellent system for keeping track of students, the interface does not automatically pull up students' class (e.g., freshman and sophomore) and major. This information is obtained from the roster site. Downloading both sets of information into a Microsoft Excel spreadsheet allows combining them, so the faculty member can have all four pieces of information (class rank, major, name, and GPA) in the same row in the spreadsheet.
(b) At any time during the semester, the faculty member can sort the students by the grade assigned, create a list of students with As and Bs, and select cut-off grades commensurate with what is needed for admission into the mentor program. The faculty member can then email the relevant students, requesting permission to nominate them for the mentor program. Letters can be sent individually to the selected students through the learning platform or the roster site. An important consideration is that if letters are sent to the students during the semester when they are in the faculty member's class, most of those invited will respond with a "yes" or a "no." The letters can be very simple, pointing out that based on the students' current (or final) grade in the class, the faculty member would like to nominate them as new mentors for the next fall semester. If a student agrees, the faculty member will nominate the student, who will be contacted within the next month by the advising office to request information to complete the nomination.

An additional column can be added to the Excel spreadsheet to indicate whether students said "yes" or "no." This allows the faculty member to send an additional invitation to students who did not respond initially, perhaps because the invitation got buried in their inboxes while they studied for exams. This measure also allows sending follow-up emails after the semester is over when the students have more time to consider their schedule for the next semester.

(c) At some point after the end of the semester, the faculty member can sort the spreadsheet by the students who responded with a "yes" and send the information to the advising office, which will send a follow-up email requesting additional information for the review process.

It is worth repeating that in larger classes with hundreds of students, it is not uncommon to nominate 30 or so students. In fact, it is necessary because only a limited number of students will be able to fit mentoring into their schedules in that next fall semester, so it is necessary to nominate more students than can participate. It is also worth emphasizing that the initial contact between the faculty and the students seems to make a real difference. In some semesters, we have used the grade sheet from one of the larger classes without faculty involvement, and the advising office sent invitations to all the majors who had As, but the response from the students was not as robust.

Evaluation and Acceptance Process

In recruiting students to be new mentors, the goal is to identify new student mentors who have a natural concern for other students or appear to have the potential to develop that ability as part of the instructional program. That said, we have been surprised at the number of times students who initially showed subdued interest but were admitted to the program changed and became very competent at mentorship. Indeed, some become so enthused that they want to return as emeritus mentors and specifically request to interact with first-semester freshmen again rather than take on the organizational role of emeriti mentors. It appears the unfolding of the program during the semester changes some students in regard to holding a more altruistic perspective than they initially exhibited.

Once the students are nominated, they are contacted by email by the advising office, seeking additional information to support their nomination. Nominated students are requested to submit the following information by email or through the online application:

(a) *Attached résumé with the names of two people who can provide letters of support*—This informs us about the student's background and ability to prepare a résumé, while the requirement for the names of referees makes the students take the application process more seriously and eliminates students who do not take the program seriously. In addition, if any of the mentor applicants have not prepared a résumé at this point, this prompts them to do so.
(b) *Student identification number*—This confirms the student's identity, particularly for those with identical names.
(c) *Email address*—This provides a means to contact the student.
(d) *Phone number*—This alternate means of contact comes in very handy because some students let their email inboxes become too full. We ask for phone numbers that will give us access to the applicants throughout the next year rather than phone numbers to temporary lodgings. Phone numbers are also quite handy near the start of the semester when we have not heard back from students,

and a plethora of emails has filled their inboxes. The permanent phone number allows us to confirm that the students are still part of the mentor program and to contact accepted alternative mentors. Key information can be provided about important dates for the program.

(e) *Major*—Majors are important when pairing groups of first-semester freshmen with mentors who have interests similar to their own. The first-semester freshmen appreciate having a mentor with a similar major or interest.

(f) *GPA*—GPA provides a cut-off point of a B average for accepting students into the mentoring program. Occasionally, a conditional exception is made if a student is close to the GPA cut-off, comes highly recommended, and demonstrates a sincere interest in the program. The student will be formally accepted to the program depending on the GPA at the end of the spring semester.

(g) *Availability preference for the days and times of both the mentor class and freshmen seminar class to meet with the freshmen students*—The mentors are required to attend the large lecture section as the class meets for the first 10 weeks of the semester. The mentors do not enroll in the freshmen seminar class, but their attendance is required because this is when they interact with the first-semester freshmen in small cohort groups. The application requests their availability for both the mentor and freshmen seminar classes. If their course registration overlaps with another class, this effectively eliminates the students from being mentors. We usually schedule the two required courses late in the day to avoid possible course conflicts and keep the mentors on track for graduation. Once all the information has been collected from the nominated students, the selection committee reviews the material submitted by the nominees.

It is useful to have a waiting list of new mentors drawn from students who were interested but had a course time conflict because some accepted mentors may decline acceptance over the summer. Students might get offered a research experience or internship, become ill, or move with their families. Often for unanticipated reasons, a number of students cannot remain part of the mentor program, and we usually are informed of this at the last minute before the start of the semester. Just as some students had initially assured us that they were available to accept the mentor position, others who were not due to a course time conflict might also change their schedule and become available to accept the position. The waiting list of students is useful to rapidly identify competent new mentors. Notices are sent out to the successful nominees congratulating them on their acceptance. Letters are also sent to inform the students who remain on the waiting list and the students who are not eligible for the program. The letters also include information about other available opportunities within the department and on campus.

Training and Engagement to Remind the New Mentors of Their Important Role

After evaluation and acceptance are completed, the program starts with an orientation near the end of the spring semester to welcome the new mentors, who will start in the fall semester, to provide an overview of what they will learn about mentorship and leadership, and to reinforce their important role in helping first-semester freshmen and advancing their own career development. This meeting is typically scheduled near the end of the traditional school day, again to reduce the chances of conflicting with students' courses and to get as many new mentors as possible to attend this optional event. The emeriti mentors are also invited.

At this meeting, where sandwiches, water, and tea are served, we invite a couple mentors from the previous year and emeriti mentors to speak to help convey the enthusiasm among the previous year's class of mentors. We also go over the broad outlines of the course and the important impacts the new mentors will have on the lives of some 800 first-semester freshmen. We know that the new mentors will not remember much of what we tell them, but we want to leave them with the positive results of what will be accomplished, their potential for important impacts, and some idea of how we will grade the leadership class. Before this meeting, we pair the new mentors with their assigned emeriti mentors, so they can experience the development of the comradery they will have with their group of first-semester mentors.

This optional meeting lasts approximately three hours, and some mentors come and go due to other course and time commitments. To force all the mentors to be present, even for an important event teaching about procedures, would change the tone of the experience. This is an important consideration as forcing their presence, no matter what else is going on in their semester, would change the experience, perhaps even from one of pleasure to one of burden. The faculty might think this situation does not give as much control as needed and could lead to problems in the preparation of the new mentors. Here, though, is the challenge for the faculty in letting go of control for the program to work. Much more is gained with this flexibility than is lost. There is enough overlap when the semester starts that another new mentor or emeriti mentor can step in to fill a gap in preparation for the first couple of weeks. Moreover, setting this relaxed tone is important to build enthusiasm in the new mentors. At the end of the day, it is the enthusiasm of the mentors that makes a real difference and suffuses the effort. April is four months away from the start of the fall semester. How much detail can the new mentors really retain from this three-hour meeting? We do show the new mentors that we care about them and provide a foundation for what we hope will become a crescendo of enthusiasm as the program builds. This optional meeting is important, but some faculty want to eliminate it and think it is not important because it is optional. Nothing could be more wrong. This is a cornerstone, without which the program may topple—although that is not immediately obvious to some faculty.

Approximately 75% of the new mentors and 50% of the emeriti mentors attend this optional meeting. This rate is because the meeting is scheduled near the end of the academic day when fewer students have time conflicts with other classes. Also, we announce the meeting approximately three weeks in advance, so the students can adjust the more flexible parts of their schedule. Finally, we order in meals (including vegetarian options) and water and tea to restore the students' glucose levels, improving their brain function so they can better retain what we tell them.

A Pre-semester Meeting

The pre-semester meeting is also optional because it is two or three days before the semester starts, and it is important to be flexible if the mentors cannot attend. In some semesters, the freshmen seminar class occurs before the official start date for the mentor course. The new mentors need to receive information to prepare for the first freshmen seminar class during the pre-semester meeting. This meeting builds on the information we have sent the emeriti and new mentors by email and the course technology.

This is a shorter meeting, lasting about one hour, and its principal focus is to pair new mentors with emeriti mentors as an additional information source for the new mentors throughout the week. This mirrors the community development that we anticipate will be cultivated between the new mentors and the first-semester freshmen. This meeting also gets the new mentors ready for their first interactions with their group of first-semester freshmen. In the first session of the freshmen seminar class, the faculty make a few introductory comments to the first-semester freshman class, and then the new mentors, who are seated with their first-semester freshmen, take over in their smaller groups. On the first day of the freshmen seminar class, each cohort group contains approximately 19 freshmen, two new mentors, and one emeritus mentor. Emeriti mentors are not required to attend freshmen seminar classes, but we encourage them to attend the first session to support the new mentors in their first experience with the freshmen.

The purpose of the first activity is to engage the first-semester freshman in interacting with the new mentors, which we do with icebreaking activities. During the optional pre-semester meeting two days before the semester starts, we present a variety of icebreaking activities and encourage the new mentors to select several in preparation for their first day with the freshmen. The new mentors should also collect the freshmen's contact information and share their own memories of being first-semester freshmen. The new mentors have an attendance sheet for the students to sign at the first class. The first-semester freshmen have a week to drop and add classes and to become familiar with the course technology platform (i.e., Blackboard, Angel, Canvas, etc.). The students should sign their complete signature, not merely initial the form, for accuracy and security on the first day of class. In future weeks, the freshmen's attendance will be recorded through the use of a remote clicker device required for several first-semester freshmen courses.

Each new mentor is given their own subgroup of 9–10 first-semester freshmen. While we encourage the emeriti mentors to attend on the first day of the freshmen seminar, it is optional. If they do attend, they split their time between the two groups of freshmen or spend more time with the new mentor who seems to need the most help. In addition, we have a few unassigned emeriti mentors available if more help is needed or if other emeriti mentors cannot attend the mentor class. During the pre-semester class, we do not discuss the grading system for the mentor course. Instead, we save that information for the first official mentor course of the semester. The grading criteria for the mentor class are available in the syllabus section on the technology platform. Typically, at least one if not both of the freshmen seminar classes occur at the beginning of the semester before the first leadership class. This is the principal reason we hold the optional one-hour session two or three days in advance of the semester for the new mentors.

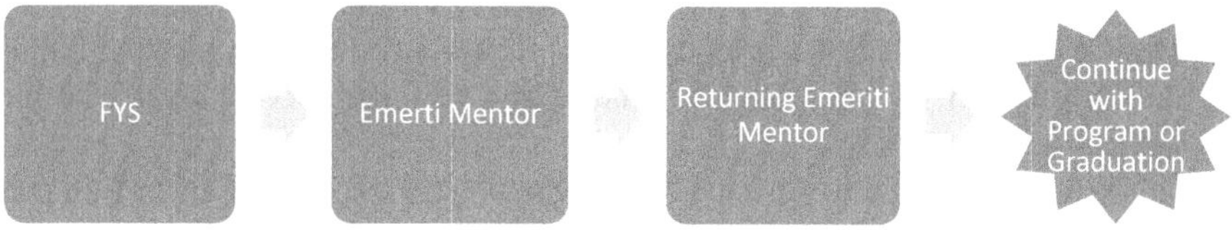

Fig. 7.1 Provides an overview of the progression of the mentor selection process as well as subsequent steps of developing emeriti mentors leading to graduation

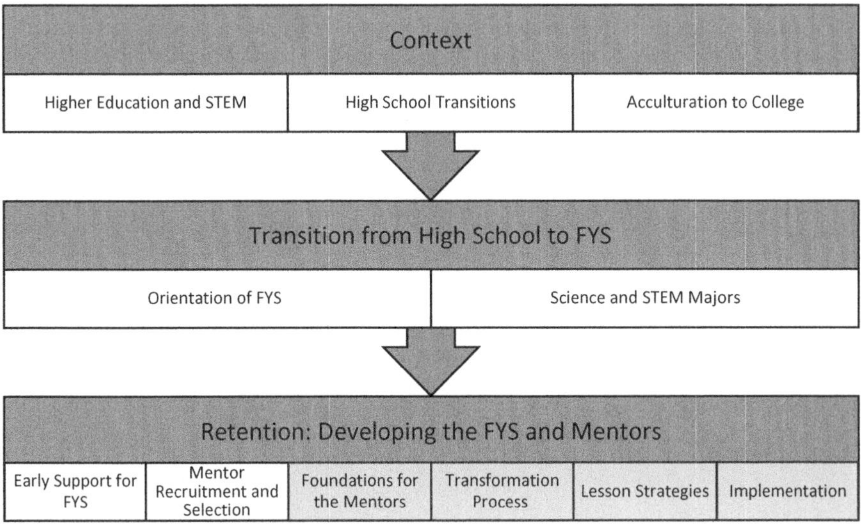

Fig. 7.2 Shows the progression in the retention process

Chapter 8
Foundations for Guiding the Mentors

Preparation for the Mentor Experience

The mentor class is designed to have the greatest impact on the freshmen while giving the mentors just-in-time training regarding what they will be involved in during the coming week. The fall semester is the ideal time for the freshmen seminar course because most freshmen are just starting at the university, and this semester has the largest number of incoming freshmen. The mentors are required to enroll in the mentor course but also must be able to attend the freshmen seminar class. The purpose is for the mentors to be exposed to the practical laboratory experience when the mentors engage in what they have been trained in the previous week. The first meeting is the pre-semester meeting two days before the semester, where we provide information about icebreakers and give the mentors time to engage in role playing with their emeriti mentors. The next week, they attend and practice in the freshmen seminar class. A variety of icebreakers are selected and practiced through role playing in the mentor group. The mentors are encouraged to initiate icebreakers in the first meeting. They can select from several options listed in Appendix A.

During the mentor class the mentors are provided a shirt that shows that they are a member of the mentor program and a list of their assigned mentees, some logistical items are discussed during this meeting. These items take up most of the pre-semester meeting time. In addition, one important aspect of this meeting is to evoke excitement about the year, the freshmen, and the overall experience. The mechanism used to engage the mentors is a form of transformational leadership. We model transformational leadership and inspirational vision in the first meeting and continue modeling mentoring characteristics and leadership styles during the remainder of the mentor course.

Some of the other items presented during the pre-semester meeting for the mentors are the use of social media, such as Facebook and Twitter. Some mentors use these as an additional way to communicate with their group of freshmen. We provide the university technology platform (i.e., Blackboard, Canvas) for the mentors

C. Harrell, D. G. Capco, *The STEM Pathway and Student Retention*,
https://doi.org/10.1007/978-3-030-65865-6_8

to communicate with and build community within their groups. The course is designed to be just in time to conserve the mentors' time and give them an immediate experience using the information that they learned in the previous week. Much of this information is discussed in the previous chapter but is important to understand how the mentors are guided for this experience.

Leadership

The first two lectures of the mentor course are intended to get the mentors organized for the freshmen in the first-semester seminar class. The mentor class focuses on leadership, which provides guidance for the mentor experience. The mentor process happens in real time. The mentors learn the content about leadership and mentoring while applying the information with the freshmen during the freshmen seminar class. This process provides a time-efficient method for training the mentors. Some programs give a semester of training before the mentors interact with the freshmen. Consequently, they might forget what they learned and not have practice putting the training in long-term memory. The training class for the mentors, therefore, is taught in the same semester as the freshmen course. This system is efficient and gives the mentors training followed by in-classroom mentoring experience within a week.

In the first lecture, it is important that the mentors interact with the freshmen as soon as possible. The mentors are taught icebreaking activities in the mentor class, and then in the freshmen seminar class, they, along with the freshmen, have the opportunity to experience these activities. In a sense, this is a lecture and lab scenario. The principal expectation of the first freshmen seminar class (lab) period is to begin to develop of a sense of community. In the second mentor class, the mentors are taught about goal planning and time management. In the next freshmen seminar class (lab), the mentors aid the freshmen in developing their own goals for the semester related to their career interests plans. In week three, the mentors receive more distinct exercises that have a greater focus on leadership. The mentors develop a vocabulary about leadership and the cohort experience. Also, in the mentor class, they learn about campus resources through activities and a quiz.

Active Learning

The researchers presented in this section have proposed active learning as an approach to addressing issues in education. Many have expressed concern about the quality of science education in the United States, as documented in *A Nation at Risk: The Imperative for Reform* (1983), prepared by the U.S. Department of Education. Concerns have also been expressed by organizations such as the American Association of Medical Colleges (1984) in *Physicians for the 21st Century, GPEP Report Association of Medical College* and the National Research

Council (2003) in *BIO2010: Transforming Undergraduate Education for Future Research Biologists*. Moreover, attention has been called to the need to make education reform based on evidence from research. For example, the 2001 No Child Left Behind Act mandates that federally funded programs be based on scientific research. The 2008 Education Science Reform Act also describes the need for rigorous research in education. Researchers have found that students learn best when they are engaged in the learning experience (Borda et al., 2020).

The characteristics of active learning include problem-based or case study learning, cooperative/collaborative learning, think-pair-share or peer instruction, conceptual change strategies, inquiry-based learning, discovery learning, and technology-enhanced learning (Michael, 2006; Michael & Modell, 2003; Wang, 2020). Active learning engages students in a way that causes them to reflect on ideas and how they use those ideas. It can be defined as "seeking new information, organizing it in a way that is meaningful, and having the chance to explain it to others" (Allen & Tanner, 2005, p. 262). Students obtain knowledge by participating and contributing. This is thought to keep students physically and mentally active in thinking and problem solving. The approach is often defined as student-centered learning in contrast to the teacher-centered learning in the classroom (Pedersen & Liu, 2003; Barr & Tagg, 1995; Borda et al., 2020).

Several studies have supported the benefits of active learning, including increased information retention and problem-solving skills, and it is speculated that these practices lead to higher confidence in the content information (Bransford et al., 1999; Wang, 2020; Colomer et al., 2020). In the sciences, discovery- and inquiry-based learning have served as important active-learning approaches. In part, this is because these fields are engaged in direct experimentation in the laboratory or the field. Students thus are presented with unsolved problems and directed to develop and test hypotheses as a natural part of the scientific process (Svinicki, 1998). Peer collaboration in the study of topics such as photosynthesis (the conversion of sunlight into chemical energy) has also been shown to result in higher performance than that achieved by students working alone (Lumpe & Staver, 1995). Peer tutoring is another strategy supported by the educational research literature (Lake, 1999). Problem-based learning is used by medical schools around the world (Correa, Pinto, & Rendas, 2003; Mierson, 1998). Computer-based learning approaches have also been shown to be successful in facilitating problem-based learning (Rawson & Quinlan, 2002; Wang, 2020). Active learning techniques have reportedly improved student attitudes and performance in science, including increasing interest in course material and class attendance and participation (Armbruster, Patel, Johnson, & Weiss, 2009; Borda et al., 2020).

In addition to medical-related areas, research has shown other benefits of using active learning. In one study, Jacob et al. (2016) found that active problem-based learning increased students' understanding of pharmacy practices. Another study on medical students, from 1989 to 1998, found that students were more engaged with curricula using active learning techniques, leading to better academic performance and lower dropout rates (Schmidt, Cohen-Schotanus, & Arends, 2009). Active learning has also been validated to have benefits for student engagement and critical thinking in the health professions, including optometry, medicine, dentistry, and physical therapy (Denial, Schreiner, Kammer, & Kim, 2015). Despite these benefits, research has shown that active learning is a transitional process for instructors, which includes developing a new vision and additional guidelines and reorganizing course material (Armbruster et al., 2009; Spronken-Smith & Harland, 2009; Wang, 2020; Bowman et al., 2019). One approach we use in the mentor class is to employ active learning.

Connecting Active Learning to the Mentor Experience

Several different approaches have been reported to be mechanisms of active learning: student responsibility, dynamic generative learning, an authentic context, learning through collaborative groups, and time for reflection. One approach that encompasses these characteristics is problem-based learning, and studies have shown that authentically crafted or situational problems provide challenges stimulating to the collaborative groups involved in problem-based learning (Boud and Higher Education Research and Development Society of Australasia, 1985; Boud & Feletti, 1991; Grabinger, Dunlap, & Duffield, 2011; Savery & Duffy, 1995; Bowman et al., 2019). Another related approach is discovery- or inquiry-based learning, as described in the previous section (Michael, 2006). These approaches are all important for the mentoring process.

Another part of the mentoring process is student responsibility. Grabinger et al. (2011) define student responsibility as "using self-directed learning activities, students are encouraged to identify their own deficiencies, to ask questions about their knowledge base, and to select and manage their learning activities" (p. 7). This information reinforces the importance of students asking their own questions about the content. At the beginning of the orientation process, we present the mentors with what is commonly called the Socratic Method, addressing topics such as what it means to be a life-sciences student. Additionally, in the mentor and the freshmen seminar classes, we continue with this topic of defining a life-sciences student by having the students consider what it means to have a career in the sciences. We also introduce the mentors to different leadership theories and encourage them to think about which they should employ to discover what it means to be a mentor for life-sciences students. Mentoring is the tool used to develop their skills and interact with the freshmen. Several activities help the mentors learn about themselves and then guide the freshmen to consider similar questions. Log forms are one of several activities that help the mentors assess their knowledge about themselves and the freshmen. This is all part of the process that provides guidance for both the mentors and the freshmen throughout this experience.

Dynamic, generative learning is "an act of active construction on the part of the student using authentic cognitive and physical processes" (Grabinger et al., 2011, p. 7). This concept is explored in this program through the idea that the mentor class is the lecture, and the freshmen seminar class is the mentors' lab. The theories they learn about leadership, mentoring, and discipline-related issues are applied in the freshmen lab seminar class. They process this information and then, through activities, are guided to implement and reflect on what they are learning. The mentors reflect on what the freshmen are learning and the outcomes of their experiences with the freshmen. The mentors process theory and their own goals and periodically check-in with the freshmen regarding their progress in meeting their goals. An authentic context is also a way to employ active learning.

Authentic Context

Authentic contexts give students experiences with encounters that happen outside the formal education system, so they experience complex tools and interactions with people (Williams & Dodge, 1992). This is important for at least three reasons. First, realistic problems are relevant to students' experiences because they address the problems and goals that students see. Second, authentic context can lead to the students developing deeper, richer knowledge structures, resulting in the transfer of knowledge to new situations. Third, team experiences with problem-based learning enables students to refine their ideas and help each other understand the content.

Historically, the purpose of education has been to assist individuals in becoming productive members of society, and the mentor experience provides students with real issues to strength problem solving skills that are needed to be contributors to society. Earlier, we described the mentor class as the lecture and the freshmen seminar class as the lab. The mentors have an opportunity to assist the freshmen with the real-life experiences of adapting to college. This assistance may include topics such as study skills, test-taking strategies, time management, homesickness, and other common challenges for first-semester freshmen. The mentors help students who decide to change their majors, consider careers other than medical school, and experience changes in social groups, such as conflicts with high school friends or roommates and the lack of a social group as well as family and financial challenges. These topics are explained in more detail in other chapters. This experience provides a context that is authentic and not simply generated scenarios from a book. There are two layers of these authentic experiences. First, the mentors analyze the problem-based challenges they experienced as freshmen, so they clearly experience this authenticity. The second layer is the authentic experience of being mentors. The mentors take information from the mentor class and create greater change in the first-semester freshmen.

Collaboration

The students need the ability to appreciate different perspectives and student experiences and to learn cooperative skills while working as a group (Grabinger et al., 2011). When freshmen register for classes, they are placed into cohort groups in which they remain throughout the freshmen seminar class. These freshmen groups gather weekly as part of the freshmen seminar class, and the mentors present them with challenges to solve, including goal setting, study strategies, test taking, mindset, discussions on science topics, and reviews of reading assignments. During the last third of the freshmen seminar class, the cohort groups are led by individual faculty to discuss specific science topics. In these group settings, there are more peer interactions and deeper discussions with the students about questions.

There is a second layer of collaborative activities for the mentors. The new mentors are placed in teams with emeriti mentors, starting with the pre-semester orientation and throughout the experience. In each meeting, the mentors are given a set of problems to discuss based on the weekly topic or real-life issues from the freshmen groups. All the group members express their approach to the situation, which provides multiple perspectives and enhances their ability to work as a team. Both the freshmen and the mentors, as stated in the previous section, are engaged in authentic problem solving, which enables developing the skillset of collaboration.

Reflection

Another criteria discussed by Grabinger et al. (2011) concerns reflection. Grabinger et al. (2011) stated that this criterion could be more important than any of the others. Reflection enables extrapolating experiences to new situations (Colomer et al., 2020). It is a process that requires students to take time to reflect both individually and in groups. The freshmen engage in this activity by answering questions from the faculty and the mentors, which prompts a discussion of goal-setting or science-related questions. The freshmen are required to post their written reflections on the freshmen course technology platform.

For the second part of this reflection process, the mentors are given numerous opportunities to reflect on their role as mentors. For example, the mentors read several articles and are given time to dwell on leadership. These are short vignettes that trigger a philosophical approach to addressing challenges. In addition to these non-traditional, brief readings, more traditional articles and other resources encourage the mentors to critically think about leadership and reflect on the process. In addition, we require each mentor to submit a reflection paper. This assignment is due once the mentors have had adequate time to learn about their freshmen cohort group, review the class information, and gain a foundation in the mentoring experience. Finally, in the last three weeks, the mentors collaborate with groups to give presentations about their experiences during the semester and the goals they solidified for their leadership experiences. It is hoped that the mentors can extrapolate this information to new situations. In their presentation, the mentors share insights into how they plan to apply this content and these new experiences in future plans. Reflection is an exercise that helps the mentors connect theory to practice to solve problems.

Capitalization

One feature of capitalization as discussed by Holland et al. (2012) is mentors' ability to capitalize on their learning opportunities as part of peer-mentor training. Holland et al. (2012) defined capitalization as voluntary self-development; for instance, when mentors are presented with particular information, they delve further into the subject material on their own. This has also been shown to work within the STEM fields, where the development of professional identity has been shown to create a sense of belonging. Mentoring fosters the development of professional identity. Capitalization facilitates making the most out of circumstances by voluntarily seeking opportunities that allow for professional development and growth. Voluntary activities include attending presentations, joining organizations, and networking with peers in career-relevant activities (Hunter, Laursen, & Seymour, 2007; Judge & Hurst, 2007). Holland et al. (2012) proposed that such students are more willing to act as mentors to their peers due to their earlier voluntary involvement and experiences in career-related activities.

Holland et al. (2012) suggested that such efforts involve capitalizing on targeting underrepresented students in STEM because without the presence of the mentors, they may feel less welcome and more inclined to leave. Engaging the students helps them find a home in STEM. Universities face the challenge of increasing the retention rate for underrepresented students. Faculty members, career counselors, university advisors, and peer mentors who sponsor career-related activities may positively impact the experience of underrepresented students. Faculty can incorporate more collaborative learning assignments and actively encourage students to join organized, peer-mentoring programs if for no other reason than to build professional relationships that can extend beyond specific courses. Such efforts can be "self-perpetuating; active encouragement of mentoring will likely result in a subsequent class of students who are willing to serve as peer mentors" (Holland et al., 2012, p. 351). Holland et al. (2012) noted the following limitation to their study: it considered capitalization activities in one semester and at only one time point, making it more difficult to identify causal links. This study included 214 undergraduate students in the fields of engineering and computer science recruited from a historically black university and a historically white university. Moreover, 52% of the mentors were African American, and 59.3% were male.

Another study supported the notion that students who volunteer for additional professional activities enrich their career development (Kiyama, Luca, Raucci, & Crumb-Owens, 2014). Not only do they gain additional knowledge and skills in academic subjects, but they also gain practical competence in intra- and interpersonal outcomes in knowledge acquisition (Wawrzynski, Jessup-Anger, & Yao, 2011; Williams, 2011). Kiyama et al. (2014) studied students employed as mentors in a summer bridge program. In the program, these peer mentors were viewed as role models and provided other students with support and guidance. The mentors responded to online surveys, created narrative texts, and participated in focus-group interviews in two phases of data collection. The analysis demonstrated that the mentors felt they and the students they mentored had an increased sense of belonging. Their personal and professional skillset and, specifically, skills development increased along with diversity and social justice. The mentors also developed a comprehensive understanding of the context of education and retention through reading of articles that introduced them to retention language theories and statistics. This led the mentors to employ their knowledge beyond the scope of the summer program and to recognize the power of individual agency. The mentors considered their experiences beyond their own academic success and retention and were committed to giving back to the community to foster the success of other students. In a sense, the mentors reshaped the environment for themselves and others and became confident about their educational journey (Kiyama et al., 2014).

In this chapter, we explained how we provide guidance to facilitate change in the students and mentors. This guidance gives the students a solid foundation in science regardless of their chosen career paths. With the institution and department mission, there is a need to provide opportunities for leadership, development, and support to meet the goals of the institution and education in general. As described in this chapter, leadership, reflection, and capitalization help the students become excited and confident in their abilities to succeed. The mentors return as emeriti mentors because they enjoy the experience. Many of them become very enthused by the mentoring process and how it aids them and want to remain associated with the program. This is possible because often first-semester sophomores are selected as new mentors so that in their junior and senior years, they can be emeriti mentors.

The enthusiasm these emeriti mentors embody has several benefits. First, their enthusiasm lights the fire under new mentors, making everyone want to do a better job. Second, the emeriti mentors often come up with new ways to improve the program, and their enthusiasm brings out the new mentors' creativity. Third, a number of elements of the program become self-running as the emeriti support the goals of the program. Finally, this process creates a cohort of students (freshman, mentors, and emeriti) that progresses as a cohort though the academic levels, giving a sense of belonging and home to all the students involved. The mentors use these theoretical concepts presented in this chapter to grow individually, affect other students, and change their environment. The mentors connect theory to practice during their experience. These theories provide guidance for the mentors and the transformation process.

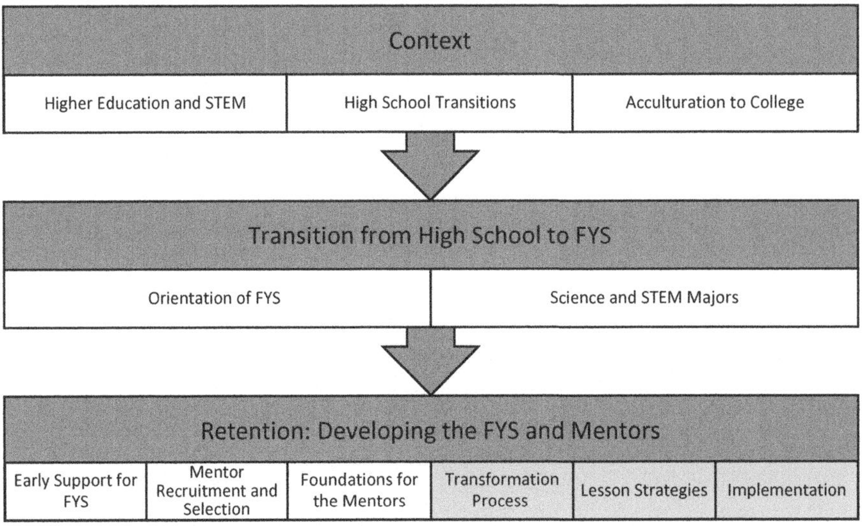

Fig. 8.1 Embodies the mentors' ways of thinking and approaching other students and shows the progression for retention

Chapter 9
The Transformation Process from a Freshman into a Sophomore

Every year of college has unique characteristics and challenges. The previous chapters dealt mostly with the freshmen year and the transitions between high school and the first year of college. In the literature, some researchers have made the argument that the senior year of college is a critical year for changes because it is the last time the university interacts with students as they transition to the real world and students' last time to expand the breadth of knowledge gained in the first three years of college (Tobolowsky, 2008; Seidman et al., 2012). Clearly, universities invest the greatest amount of resources in the freshmen year. The institutional investments in freshmen support the development of a sense of community, effective study habits, and time-management skills as students enter their sophomore year. Tinto (1993) focused on first-year attrition and, as discussed in other chapters, highlights the importance of social integration, extracurricular involvement, and institutional commitment (or students' commitment to stay at the institution as well as the institution's commitment to assist the student in being successful). When freshmen successfully adjust to college (Thomlinson-Clarke 1998; Braxton, 2019), they show higher levels of resilience when encountering challenges during their junior and senior years. Mohr, Eiche, and Sedlacek (1998) explained that seniors have different needs but are more likely to graduate if they receive needed assistance during their freshmen year. This also implies that aspects of Tinto's (1993, 2012) theory may be more important for different class levels (i.e., juniors, seniors). Tinto (1993) proposed the actionist model and discussed the importance of social and academic integration and institutional commitment in increasing student persistence from the freshmen to sophomore years.

The transition from the freshmen to sophomore years brings a different set of needs than the transition from high school to the freshmen year (Graunke, Woosley, Turner, & Thompson 2005, Seidman et al., 2012; Gregg-Jolly, et al., 2016). Sophomores need to finalize their career and personal goals during this time. Attrition from the freshmen to sophomore years can occur because students have not finalized their career decisions and personal goals. Gardner (2001) finds that confirming a major and setting career goals are sophomores' biggest challenges.

C. Harrell, D. G. Capco, *The STEM Pathway and Student Retention*,
https://doi.org/10.1007/978-3-030-65865-6_9

Some level of support during each academic year is important for student success. Gardner (2001) reports some of the unique needs of sophomores. In addition, the current sophomores are typically millennials. Pattengale and Schreiner (2000a, b) observed that sophomores often have less opportunities for student leadership, less attention from the institution, and less contact with faculty.

Millennials

Millennials are individuals born between 1982 and 2002 (Rickes, 2009). Unlike previous generations, this diverse population includes different "behavioral patterns and traits" (Monaco & Martin, 2007, p. 94). Monaco and Martin (2007) and Rickes (2009) described the following seven core characteristics, patterns, and traits of the millennial generation: "sheltered, team-players, conventional behavior, confident attitude, achievers, special, and pressured" (Monaco & Martin, 2007, p. 94). These and other authors believe that these qualities lead millennial students to lack critical thinking skills and confidence in their academic skills, so they do not realize their true academic capabilities. This group of students need "immediate feedback, praise, attention, and guidance" due to great involvement by parents, sometimes referred to as "helicopter parents" (Monaco & Martin, 2007, p. 94). This behavior may be influenced by these students' exposure to digital gadgets and social media from birth, which may reduce their ability to think critically, resolve conflict, and feel comfortable with face-to-face interaction.

These developments are important because these skills are often crucial for success in college, and 75% of freshmen may not return for their sophomore year (Monaco & Martin, 2007). In addition, 60% of students may not continue in college or choose a different institution after struggling during the first year of college. If students decide to transfer to another institution, this process may cost the taxpayers, federal government, educational institutions, and students more. Often, credits do not transfer equally, causing students to take extra time to complete the degree. In addition, institutions invest financial resources, effort, and time to retain students from freshmen to sophomore year. Millennials at every academic level might be struggling with critical thinking skills, have unrealistic college expectations, and become stressed about not always receiving immediate feedback (Andrade et al., 2020; Council et al., 2020). Due to these challenges, students might ultimately not complete their degree.

Challenges During the Sophomore Year

Among millennial students, the decrease in institutional efforts during the sophomore year can especially affect those who have not yet selected their major or change their major during the freshmen or sophomore year (Tobolowsky, 2008).

Another challenge in the sophomore year is that some students may be completing remedial courses or repeating courses from the freshmen year. These students might not effectively be at the sophomore year level and in fact enter their third year of coursework before they begin the same coursework as their peers. At the same time, advance placement credits from high school might place students at the sophomore level although it is their first year at the institution.

In general, when moving to their junior year, students typically begin to take upper-division-level STEM courses (Gahagan & Hunter, 2006). The hope is that if students receive the needed developmental support during their freshmen year, then they learn "the ropes" and become acclimated to the STEM community in their sophomore year (Packard, 2004, p. 290). In this year, students may experience confusion, stress, and uncertainty regarding their choice of major. They may also question whether there is still a need for college given the money spent on college and their indecision about their career options. Moreover, the rigor of the coursework progressively increases as the students continue through the academic program. This rigor requires greater time and skill to master the content, making the energy investment in college for student success even greater. Students' mindset can even cause them to question the purpose of universities' general studies requirements, such as humanities courses, because they want to focus on what they view as their desired career path. The advanced STEM classes and the general studies courses impose greater demands on their time, resulting in pressure and stress during the sophomore year. Consequently, students question their reason for needing college and their success in college, which is unique to STEM majors (Tobolowsky, 2008; Gregg-Jolly, et al., 2016). If some of these issues are not resolved during the freshmen year, students will still need to address them during their sophomore year.

Sophomore Solutions

One solution to address these issues suggested in the literature is to provide sophomore students with opportunities to reflect on their state of understanding of their career development (Gahagan & Hunter, 2006). Most institutions do not provide a structured system that supports this process, but a few institutions have added peer-mentoring programs that include time for this structured reflection. In addition, advising offices have developed techniques for this sophomore-year reflection, described as "turning inwards and exploring how one fits into college life and the world at large," which is supported by psycho-social mentoring theory (Williams & Manning, 2014, para. 5; see also Gahagan & Hunter, 2006; Sanchez-Leguelinel, 2008). Psycho-social mentoring theory includes counseling, role modeling, and community development, as explained in previous chapters. All of these enable students to become more confident in their life, career options, happiness, and future direction.

In addition, STEM major students face unique time commitments, including the extensive formal laboratory courses required within the STEM curriculum.

Sophomores have incomplete laboratory skills, requiring them to devote a significant amount of time to learning these techniques. Even though the students learn techniques within the STEM curriculum, an advanced set of skills is also learned when working as a team in a faculty lab under faculty supervision. This experience is distinct from the lab experience with other peers within the traditional coursework. This adds to the issues in the sophomore experience and the retention issues for freshmen in STEM majors discussed in the previous chapters. The literature suggests the importance of encouraging the involvement of freshmen and sophomores in these research experiences (Micari et al., 2010; Nerio et al. 2019), but this poses a challenge due to their incomplete laboratory skillset. Along with the time commitment, this could create an additional level of stress and pressure.

Involvement in research is important so that students can be best prepared for their STEM career choice. The challenge is to connect students at the freshmen and sophomore levels to STEM research experiences. Friedrich (2014) suggested the following solution to this issue: using real-life datasets with the freshmen and sophomore classes to reinforce the scientific method, help them become excited about data and research, and connect them to real-life experiences in the STEM field. This experience provides novice STEM students with opportunities to combine training with the creation of new knowledge. The computer-based research involves minimal training time compared to laboratory-based research, so students can include it in their schedule and course content during their freshmen and sophomore years.

At the end of the academic year, freshmen are better prepared for the transition to their sophomore year as a result of the mentoring program. In our program, as discussed, the freshmen are introduced to SMART goals, time management, and other activities that assist in developing social interactions. Moreover, they learn efficient note-taking skills, which many often lack coming from high school, and about campus resources to support academic success. These activities take place during the first semester, and the freshmen can use these skills during the second semester. This process influences student persistence from the freshmen to sophomore years. Homesickness, which can be a major problem for many freshmen in the first semester, is more easily handled in the second semester because students have developed friendships and other supportive relationships with their academic cohorts (i.e., Academic Success Clusters) and peer mentors through the introduction to student clubs and other organizations at the institution.

Between the freshman and sophomore years, greater changes are often seen. This transition can be compared to learning a new sport. When a person first learns a new activity, the brain integrates the muscle coordination needed to perform that activity. The integration of the muscle memory is developed during sleep. What was a frustrating experience on the first day of effort becomes an easier experience on the second and subsequent days. The summer between the freshmen and sophomore years is a time of similar integration. When students come back as sophomores, the world is, and therefore college is, different to them. The mentor program supports the development of resilience to return for the sophomore year. Again, the mentor program provides cohorts, freshmen seminar classes, advising, engaged faculty and staff, and social integration.

In our program, we assist the freshmen in preparing for what they may encounter their sophomore year. This includes providing opportunities for the freshmen to step into leadership roles as mentors for future freshmen or other leadership positions within the department or college in their sophomore year. We consider it important to involve sophomores in serving as new mentors in our programs. This has several advantages. It engages sophomores in social integration, provides them with opportunities to reflect on their career goals, and produces critical thinking. In addition, by recruiting sophomores who recently completed the freshmen year in our program, we facilitate the sophomores who are new mentors to think about their experiences in a new way.

It also allows two years for students to return as emeriti mentors. Students who return as emeriti mentors for one or two years are engaged and enthusiastic and work through their own persistence problems. They thus serve as enthusiastic coaches to new mentors, helping pass on the sense of belonging to first-semester freshmen. If they do not continue as emeriti mentors, they stay engaged within the department and college community, as Tinto (1993) suggests is necessary for student success.

Regarding retention, sophomores' involvement in the mentoring program gives them leadership skills and attention and prevents the isolation sophomores may experience. This aligns them with the path of an engaged learner. A group of new mentors is recruited from the freshmen seminar class, and some sophomores who enjoy the experience return as emeriti mentors in their junior and senior years. This helps solidify for both groups their career goals and directions and increases their commitment to stay at the institution. They seem to have great enthusiasm and coach the new mentors better each year. Their level of critical thinking and their understanding of campus policies increase, helping them keep on track to graduate on time. This structure, particularly the tiered support from peers, faculty, and staff, also keeps them from making mistakes in course selection.

The assignments in the leadership class provoke the students to process and think about their experiences in a new way. The students are better sophomore students and better academic citizens. These students have matured, successfully completed the first year of college, and gained skills to tackle the academic rigor of college. The mentors develop empathy and insight into the wellbeing of their mentees and recognize when freshmen encounter the typical challenges of the freshmen year. There has been discussion of continuing the mentor program and cohorts into the second semester of the freshmen year or even into the sophomore year, but this has not yet been implemented. These ideas are discussed further in Chap. 10.

Summary

In this chapter, we discuss the core issues freshmen experience that affect persistence and success during the sophomore year. Some core issues are career indecision, social integration, institutional commitment, career reflection, STEM time commitment, remedial coursework, and laboratory-skills-development experiences. The sophomore year is the time when many institutions provide the least amount of support and instead focus on the next group of freshmen entering the institution. Sophomores have fewer interactions with faculty and less programmatic support. Institutions should reconsider offering additional opportunities to better support the sophomore-year experience. Within this chapter, the suggested solutions included expanding the mentor program and using computer data to develop sophomores' research experience. Every year, we continue to adapt the program to give the mentors and emeriti mentors a better experience, and we view this as aligning individual steps in a journey. To reach perfection is unrealistic because developing a mentor program and creating a positive student experience is a continual process. A highlight of our program is the number of STEM students who return for additional semesters as emeriti mentors. A successful mentor program hopefully creates a change in freshmen between the first and second semesters and between the freshmen and sophomore years. College is all about setting students up for success, and this entails providing them with safe opportunities to make mistakes, change their behavior, learn, and grow.

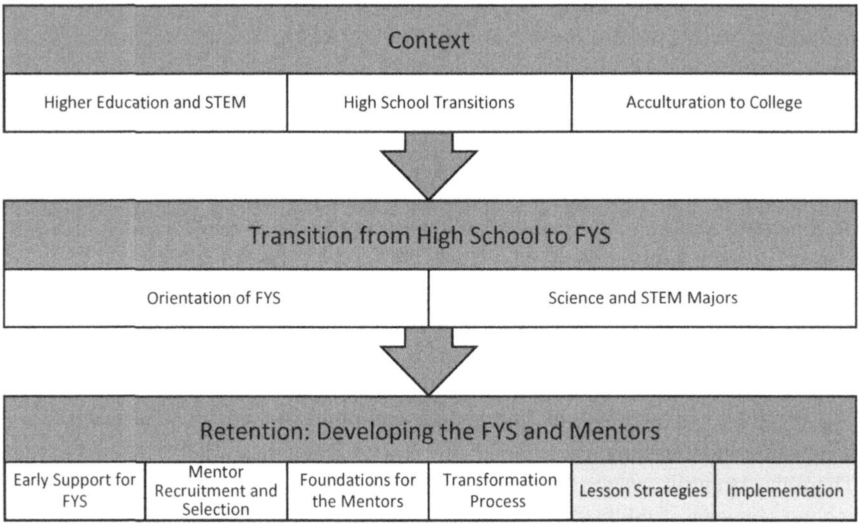

Fig. 9.1 Summarizes the process with a focus on the freshmen to sophomore transition

Chapter 10
Lesson Strategies

Introduction

In this chapter, we discuss lessons learned from our involvement with mentoring programs to address some challenges of meeting the goals of multiple stakeholders, who include the students, faculty, administration, government, and community. The challenges include the growth of student numbers, eliciting ideas from faculty and administrators, the changing student population, and achieving collaboration among stakeholders. Stakeholders have different ideas about the role of the freshmen seminar class and the mentors. The purpose of the STEM freshmen-year experience is to increase retention through social integration, academic skill development, the teaching of discipline-specific concepts, and the formation of a holistic person. These factors influence graduation rates and ultimately student success (Seidman et al., 2012).

History and Development of the Mentor Program

As stated, institutions are evaluated on retention and graduation rates, and often these numbers are low in STEM fields (Seidman et al., 2012; Spaulding, Kennedy, Rózsavölgyi, & Colón, 2020). The goals of all the stakeholders are to increase freshmen retention and graduation rates. Researchers report that these numbers increase when students feel a sense of belonging to the institution and department (Strayhorn, 2012, 2015; Atkins et al., 2020).

In agreement with Strayhorn (2012, 2015), surveys and institutional data showed that STEM students often did not feel connected to the institution, providing a context for the need for mentoring programs. In discussions, mid-level administrators set goals to develop a mentor program and other first-year programs to begin strengthening students' connection to the psycho-social aspects of development in

C. Harrell, D. G. Capco, *The STEM Pathway and Student Retention*,
https://doi.org/10.1007/978-3-030-65865-6_10

the department, institution, and field. At first, a pilot program with first-semester freshmen and upper-class students was introduced. Based on its success, the program was expanded to include more freshmen each year until the third year, when all freshmen students with biology-related majors were required to participate. In the seventh year, this population included approximately 815 freshmen. As a result, the mentor program increased in size, and the organization of the program needed to be adapted to accommodate the number of students. For example, there was less time for active learning exercises during the mentor leadership class. Some activities were shifted to out-of-class assignments with brief discussions during class. This occurred because with larger numbers of students, more time was needed to address questions or issues related to mentoring the freshmen. It was also important to engage a large group during class instruction. Every year, the approaches used to interact with the freshmen have been adapted based on feedback from administration, faculty, mentors, and freshmen.

Administration Perspective

The administration focuses on freshmen retention and graduation rates. When working with other stakeholders, this is the lens that guides decision making. At times, the administration has a vision for the end goals and the process but may not see the practical implementation of the goals on a day-to-day basis. This view also impacts the development of programs to support reaching the desired goals. This is the benefit of having several stakeholders participating in the discussion to assist with implementing the vision. There is a need for clear communication between higher administration and those implementing the vision, including faculty and staff. The joint development of a hierarchy of goals agreed upon by faculty and staff brings the most rapid progress. For example, during the original development of the mentor program, the staff, administration, faculty, instructors, mentors, and freshmen all provided input and assisted with the development and evolution of the program.

Faculty Perspective

The perspective of the faculty members may depend on how much they understand the integration (i.e., transition from high school to college) of the freshmen experience. The faculty members are influenced by their academic rank and experiences. New faculty members focused on tenure may have a less global perspective of student development and the concept of freshmen retention. The high school experience has changed over the decades, and even senior faculty may not understand the amount of high school preparation provided for subsequent college. They may overestimate the breadth of knowledge possessed by the incoming freshmen (Osa et al., 2013).

In addition, the faculty members may have different expectations about the first-semester experience than the freshmen themselves. Midlevel and senior faculty may have a more comprehensive view of what the freshmen-year experience is and should be. Accordingly, the faculty have different perspectives on the freshmen seminar class and overall experiences. For example, some faculty may think that a content area should be covered in detail during the freshmen seminar class, while other faculty may favor broader exposure to science, and others might still be interested in critical thinking. Fewer faculty members understand the concept of socialization and the integration of the student transition from high school to college. Depending on faculty members' experience, they may have differing concerns about the outcomes for freshmen, and only the most vocal faculty may share feedback with the administration. It is important for faculty and administration to work together (Iverson, Bragg, & Eddy, 2020). Providing a foundation for freshmen retention may help the faculty assist the freshmen with the transition to the first year of college. Some faculty would like the seminar class to provide a head-start to understanding the STEM content material. Other faculty view the course as one ancillary to other active learning courses. The seminar class provides a foundation for STEM and the first-year experience so that students can be successful in other introductory STEM courses. Priorities in the seminar class include social integration and academic skills development, with content learning as a secondary objective. This does not mean that every portion of the class focuses on the STEM content and reasoning, but rather that examples are pulled from STEM content that are meaningful for a first-semester freshman.

The faculty might also struggle to understand the role of the mentors. One pitfall in mentoring noted earlier is that the faculty are often uncertain about how to involve the mentors in class because they did not have this experience in their training. This can easily lead the faculty to treat the mentors as teaching assistants. This is not advisable because the roles of teaching assistants and mentors differ. Mentors are more engaged in the development and integration of social activities centered around science, whereas teaching assistants have a greater level of involvement in grading papers, tests, and other assignments. Often times, the teaching assistants do not perform their duties in the presence of freshmen, while the mentors interact with them inside and outside the classroom as their principal role. If the faculty members tend to use the mentors to, for example, hand out or collect papers, then the faculty are not using the mentors correctly. Departments have many astute and brilliant faculty members, but that does not make them the ideal individuals to engage the first-semester freshmen. Knowledge in the STEM content areas does not mean that an individual is relatable and engaging to millennials.

In addition to knowing one's content area, there are different skillsets to engage the freshmen population. Identifying these faculty members may be challenging but is possible. Having faculty in a diversity of research areas is beneficial because it introduces freshmen to alternate STEM career paths and disciplines. In addition, freshman learn the hierarchical organization for STEM disciplines. This factor is even more important when connecting with minority students and developing cultural competence.

Mentors Perspective

The pool of the students who become mentors often falls into two categories: students who did not need a freshmen seminar class and students who gained a great deal from the experience. Both of these categories of mentors value the learning experience of leadership and the socialization process. The first category rapidly learns the importance of the seminar through the instruction provided in the leadership class. The mentors who participate in the experience can see the immediate goal in helping the community of the institution. Both categories of mentors strive to be open in the mentoring experience.

There are several benefits of the mentor experience. It allows the mentors to connect theory to practice, which is beneficial for letters of recommendations in their future career. Their growth during the experience can be clearly commented on in their job applications. Another key aspect of this experience is the time for reflection and internalization of this process, through which the students gain insights and can observe their growth. Overall, the mentors consistently comment in reflections, class discussion, presentations, and evaluations on how this experience has impacted them and how they enjoy helping others. This experience reminds them of their own growth since their first semester of college. It reinforces the tools they will need when beginning the next phase of their career. For example, a mentor commented that they forgot the importance of having dinner with people daily for first-semester freshmen.

The freshmen's observations confirm for the mentors that they have grown as students. The experience also reminds them to use this information during the multiple steps of their career and increases their employability because they can better understand their own learning process. They are able to provide examples of their growth experiences and have the tools to succeed. They have grown in their leadership and communication skills. They feel a part of the community because they know individuals in the department and have been forced to learn or find resources based on questions from their mentees. They want more time with mentees and additional options for rewarding them for their mentor interactions (e.g., points or extra credit). The mentors also share feedback from extending the experience into the second semester and the sophomore year. In addition, they want to have a transfer mentor experience. The mentors often return each year due to the sense of belonging and important work in helping the STEM community. They may return in different roles, such emeriti mentors or project, planning, or research assistants within the program.

Freshmen Perspective

Developmentally, some freshmen might not initially recognize the benefits of the first-semester seminar course. After the results of the first exam, many freshmen better appreciate the tools, resources, mentors, faculty, and administrators available to them through this course. Points in the class encourage mandatory participation in study groups, which the freshmen might not initially think of as important. The purpose of the experience is to help them learn the characteristics and behaviors of successful STEM students, and this is not always obvious to freshmen at the beginning of the year. Their perspective is based on their experiences from high school. They are learning what it means to be a college student. They begin their interactions in the course from several perspectives, including large group interactions, individual breakout and mentor interactions, and advising resources. Initially, the freshmen might not recognize the benefits of the course but often enjoy interacting with the mentors.

Expectations from the Government and Community

Current issues in the government and community relate to the concepts presented in Chap. 1 and the reasons why a mentor program is beneficial. The issues include increasing financial costs; pressures for increased retention and graduation rates; the demand for skilled individuals, especially in STEM; and the need for a diverse workforce internationally and in the United States. These are expectations from the government and community that educational institutions are expected to meet. A mentor experience provides one avenue that is beneficial for multiple stakeholders and addresses several of these issues.

Programmatic Lessons

Our program experienced several growing pains as it increased in size. One challenge successfully met was the use of technology for communication, organization, record keeping, grading, and community building. Technology provided rapid analysis, including surveys and the determination of grades for attendance, outside class activities, extra credit, and discussion board questions. To develop a program, it is recommended to use an existing platform to organize and streamline information. There are several educational platforms that can assist with this process, such as Canvas, Blackboard, Angel, LoudCloud, and Turning Point.

When starting a new program, as with any group activity, the larger the group is, the more suggestions there are for what should be done. Experience suggests that input has value, but attempting to apply all the input from all stakeholders at once falls under the scenario of too many cooks in the kitchen. This scenario might occur with an individual charged with implementing a freshmen-retention initiative. We suggest that anyone implementing a new program examine the current research and theoretical models on retention and select proven approaches, such the recommendations described in this book. It is also important to utilize the existing technology already in use at the institution (i.e., Blackboard, Canvas, and Turning Point). As a new program grows, suggestions from other stakeholders and published data from the literature can be added to expand the retention initiatives. It is recommended that new ideas be piloted with a smaller subset for first-semester freshmen rather than applied to all entering STEM freshmen. If a new idea creates an aberration during this process, the impact is minimized, and the pilot group can be modified for improvements.

For existing freshmen-retention initiatives, new faculty and administrators often have ideas to improve them. In that scenario, there is also the risk of too many cooks in the kitchen. An effective retention effort often will gain the interest of faculty, staff, and administrators. Again, a gradual approach is recommended to implement the ideas for all stakeholders. The use of pilot groups and strategic planning can be highly beneficial to achieve the desired outcomes.

The number of mentors involved in the program may influence the way the mentor training is delivered. The number of mentors is linked to the number of freshmen in the program, which, in turn, may increase due to higher enrollment in the institution or the extension of the research initiative from a pilot group to all first-semester freshmen. With a smaller group of mentors, time can be provided during the mentor class for active learning, class discussion of theory, and application-based activities. In small groups, there is time to use examples from the mentors' experiences to process issues encountered in their interactions with the freshmen. As the number of mentors increases, this activity is shifted to outside-of-class assignments. For example, the mentors may watch a series of videos outside class and then proceed with a brief discussion during the class period. The use of scenarios derived from the mentors' experiences with their freshmen that week requires time to appropriately process and internalize the theories and understand how to solve the issues. With more students, there are more scenarios, so it is important to prioritize the values in the program. This is a challenge that must be successfully met as a program grows. It is a challenge to work with more students but a success to reach more students. How these concepts can be implemented is described in the next chapter.

Conclusion

The stakeholders want the students to be successful. They all have a common goal to help the students graduate and become integrated into a society that is connected to the mission of higher education. Often, the stakeholders have different perspectives on how to help the students reach these goals. Within this struggle, though, we find the greatest success for all involved.

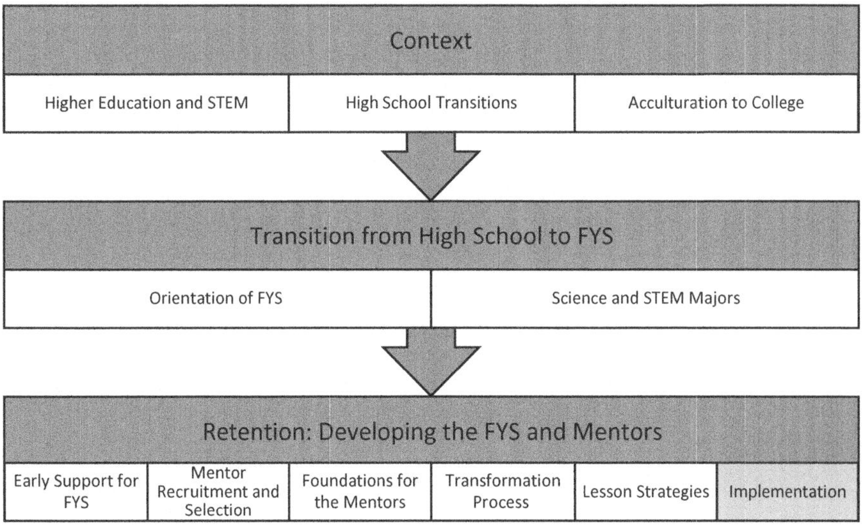

Fig. 10.1 Summarizes the process amidst the complex perspectives of the stakeholders

Chapter 11
Implementing the Plan: The Weekly Schedule

Introduction

Often, colleagues ask about the content for the freshmen seminar and mentor classes. They inquire about how to develop a program in their departments or institutions. In an attempt to address these requests, the earlier chapters have presented the approach we use to recruit and train mentors and emeriti mentors. In this chapter, we review what occurs in each week of the two linked classes, that is, the mentor class and the freshmen seminar course. The mentor class is connected to materials and websites presented in the appendices to give readers access to the training materials for the mentor class.

We have a few principal objectives. The first is to create a welcoming environment for the mentors by modeling what we want them to do with the freshmen mentees. Second, we inform the mentors of their important role in aiding the freshmen in planting the seeds for the freshmen to grow as well as to make important changes in their own lives. We continue to reinforce these principles throughout the course. We are intentional about the repetition of information and, even more importantly, time for the mentors to engage in reflection as they move forward with the experience. A great deal of information is covered, and it is necessary to make sure that the key points are reiterated to support what the mentors are experiencing during the process. The third principal objective is to establish in the freshmen a sense of belonging by promoting social interactions and to support psycho-social aspects of their development in the department, institution, and STEM fields.

In the following, we outline the activities engaged in each week in the mentor class and the freshmen seminar class. The training in the mentor class comes before the freshmen seminar class that the mentors must attend. We have found this to be an efficient way to provide just-in-time training to mentors and have the mentors experience the laboratory situation in the freshmen seminar course. Thus, the list of activities starts with Week 0, or the week before the start of the semester. We term

C. Harrell, D. G. Capco, *The STEM Pathway and Student Retention*,
https://doi.org/10.1007/978-3-030-65865-6_11

this Week 0, when only the mentors meet, to keep the sequence of weeks consistent with the semester calendar of 15 weeks.

Week 0

Mentor Class In Week 0, we hold a meeting to provide the second part of the orientation to the mentors. This meeting pairs two new mentors with an emeritus mentor, which remain an intact unit throughout the remaining 15 weeks of the semester. This activity mirrors what the first-semester freshmen will experience with the mentors in the freshmen seminar class that begins in Week 1. The meeting in Week 0 gives the new mentors and emeriti mentors ample time to reflect and prepare for Week 1.

During this meeting, we prepare the mentors for meeting the first-semester freshmen in the freshmen seminar class. Information is covered in the first mentor class, and this information is best communicated with enthusiasm, based on the discussion in Chap. 3 on the identity and biological development of nervous and anxious mentors and freshmen. In the mentor course, we have the mentors role play how to develop relationships with students through activities, such as icebreakers (see Appendix A), which help both the mentors and the freshmen get to know one another and develop a comfortable environment. We provide a context for education and discuss what makes a successful student as well as the importance of the program goals.

As the Week 0 mentor class starts, we make sure that the new mentors are physically sitting with their emeriti mentor as a visible sign that they are part of the cohort and members of several communities. This mirrors the freshmen seminar class, where the new mentors will sit with their freshmen mentee cohorts. We involve the mentors in creating a lesson plan for their first interactions with the freshmen mentees (Appendix B). We discuss the technology used in the class, such as social media (Appendix C) and the class learning management platform (Appendix D), in our case, Blackboard. There are two course-learning-management platforms, one for the mentor course and one for the freshmen seminar course. In the latter, the mentors have access to parts of Blackboard to stay in contact with the freshmen, for example, to suggest meetings outside class and to promote discussion though the platform email and announcement section. The size of our program requires that we prepare these platforms to purposefully create a welcoming environment for the freshmen and to collect and organize the scores within the freshmen seminar class.

In preparation for Week 1 of the freshmen seminar class, large signs are made so that in larger classes, the freshmen can find and congregate around the two mentors assigned to their group. The freshmen are also sent emails to remind them to look up their course line numbers and titles, so they can rapidly organize into their groups on the first day of the freshmen seminar class. While the emeriti mentors are not required to attend the freshmen seminar class, we ask any who can to volunteer to help on the first day to support the new mentors and organize the program. We also use large signs in the Week 0 mentor class to bring together the new mentors and their assigned emeriti mentors, again mirroring activities that will happen in Week 1.

Week 1

Mentor Class We briefly review some of the concepts discussed during Week 0, specifically what makes a successful student, program goals, and the technology used in the class (i.e., social media and Blackboard). In addition, we cover the syllabus and grading for the mentor class and introduce leadership theory (Appendix E) and mentoring theory (Appendices F). We prepare freshmen activities for Week 2, including student goals (Appendix G), time management (Appendix H), and campus resources, such as counseling and tutoring.

During Week 1, we teach the mentors how to facilitate group and class discussions (Appendix I) and introduce the content (Appendix J) that will be covered in the freshmen seminar class in Week 2. This content is applicable to freshmen pursuing careers in science and medicine and includes the pros and cons of the biological sciences. The content includes topics such as the level of precision in science, the foundation of science in repeatable experiments and controls, the influence of big business in biology in the twenty-first century, the influences on personal lives—referring to the time allotted to science vs. family life and the rewards and satisfaction scientists might have with the outcomes of experimentation.

Freshmen Seminar Class The mentors hold up signs so that the freshmen can sit near them in groups, and some emeriti mentors volunteer on this first day to help the freshmen find the groups. In Week 1 of the freshmen seminar class, the mentors facilitate icebreakers as a way to engage the freshmen in talking about themselves. Each mentor has 9 or 10 freshmen mentees per group within a cohort of 19 freshmen and one emeriti mentor. The instructor for the freshmen seminar class covers the syllabus and course technology. The mentors collect attendance sheets for their cohort groups, exchange contact information, and schedule a time outside class to meet with their groups. In an attempt to create a welcoming environment, the emeriti and new mentors arrange activities outside class during the course. The purpose of these activities is to allow the freshmen and the mentors to learn about one another and to drive the beginning stages of the relationship development and mentoring process.

Week 2

Mentor Class During Week 2 of the mentor course, we continue preparing the mentors to address what makes a successful student and consider possible career goals. We introduce leadership and mentoring theory and prepare the mentors to facilitate activities for Week 3, which covers SMART goals (Appendix G), time management (Appendix H), and campus resources. To assist with teaching the mentors about campus resources, they complete an online quiz. Reflection on the aforementioned information often gives the mentors a more comprehensive understanding of themselves in the process of helping and learning about other people.

Freshmen Seminar Class In the freshmen seminar class, majors and concentrations within the department are presented, giving the freshmen an overview of the degree options. An article presented to the freshmen in the previous week through Blackboard (Appendix J) is discussed to support the purpose of the course goals and discipline content. For the students who did not read the article in advance—often a large number—key points are provided to allow them to be actively involved in the small group discussion with their mentor. The small group discussions facilitated by the mentors focus on the most significant points of the article. The freshmen also discuss what they think is most important to them in the article.

Week 3

Mentor Class In this course, we begin to engage the mentors in reflecting more on their freshmen cohort. We ask the mentors how things are going with their freshmen and what they have learned about their freshmen. We discuss logistical issues with their groups of nine or ten freshmen. We talk about issues such as whether their freshmen are present in the proper mentor groups and have found all of their classes. We ask them if the freshmen know how to use Blackboard. We ask the mentors about how the freshmen are getting along with their roommates in the dorms or if they are feeling homesick. We encourage the mentors to remind the freshmen that these are normal experiences and to provide tips to help them connect to campus resources. We review the quiz content as well as goal setting, time management, and campus resources.

Freshmen Seminar Class In the freshmen seminar class, we provide a 10-minute overview of goal setting (Appendix G), time management (Appendix H), and campus resources. Then the mentors meet with their nine or ten freshmen for an additional 30 minutes for a short icebreaker and then discuss SMART goals (Appendix G). A SMART goals sheet (Appendix G) is given to each freshman to complete during the freshmen seminar class. On the sheet, the freshmen write down one academic goal, one involvement goal, and one social goal. The mentors collect the documents to discuss in the mentor class and with their freshmen in coming weeks.

Week 4

Mentor Class In the mentor class, the mentors are again asked to consider some problems experienced by the freshmen in their cohort, and other mentors and the emeriti mentors are asked to suggest solutions. We ask how their freshmen are adjusting to college and if any are feeling disoriented or homesick or struggling in their classes. This not only spurs collaborative thinking but also provides another way for the mentors to reflect on the information provided and, at least for some, to turn this into knowledge rather than simply a series of facts. This week, we provide in-depth content on popular leadership theories in the field, particularly situational, servant, and transformational leadership (Appendix E). Also, within the course, articles that the mentors should review are given to them to allow them to dwell on leadership (Appendix K). During this session, the mentors are instructed in leadership and mentoring vocabulary to apply in the freshmen seminar course. Also, we teach them how to use these concepts when they revisit SMART goals (Appendix G) in the freshman seminar class during Week 6.

Freshmen Seminar Class In the freshmen seminar class, the freshmen receive advising information. With advisors, they discuss degree planning for their majors in preparation for spring registration, which is about a month before freshmen class registration. At our institution, we discuss the importance of course completion each semester for on-time graduation. A pre-health advisor also shares information about medical school and other pre-health professions and discusses these career goals in relation to spring registration.

Week 5

Mentor Class In the mentor class, we again ask the mentors how the freshmen in their cohort groups are transitioning to college and if they have encountered any specific challenges during the first weeks of the semester. Either this week or the next, many freshmen will have their first exam, and others will receive their exam results. The mentors ask their cohort if they are concerned or surprised about their exam results. At times, the freshmen might not admit to any issues, but this question can still prompt them to think about their situation and perhaps recognize that others in the cohort are experiencing similar problems. This reflection may, for the first time, give the freshmen a reason to think that the study approaches they used for their first exam perhaps were not the best. Getting the freshmen to think about their situation (when they did not do well) aids in breaking the pattern of how they studied in high school. If they are not prompted to think about this, they may continue to rely on study habits from high school and not recognize the need to adjust their study habits to be successful in college. As stated near the beginning of this book, many students need to recognize the importance of acquiring additional study skills before they are willing to do so. In future weeks, the freshmen tend to become more willing to recognize that they need help and more open to attending tutoring and study sessions provided by the mentors and using other resources available at the university.

In Week 5 the mentors review their SMART goals (Appendix G). We give the mentors time to discuss how to facilitate the SMART goals activity effectively and hold discussions in small groups (Appendix I). Exam results can cause a culture shock for the freshmen, making them more receptive to assistance from the mentors. In this class period, time is also spent engaging the mentors about best approaches to individualize other activities and giving appropriate suggestions for resources on campus.

Freshmen Seminar Class In the seminar class, Week 5 is different than previous weeks. This typically is the first week when faculty guest speakers are invited to share about their careers and discuss research modeling approaches used in the scientific process. Also, these faculty share decisions and choices they considered to have been successful in their own careers. The speakers represent various subdisciplines within the department faculty. These can include presentations from medical doctors, alumni, pre-health professionals, and researchers who focus on ethics, sustainability, and ecosystems. There are two scientific presentations and opportunities for the freshmen to ask questions during the class.

Week 6

Mentor Class Again, the mentors are asked about the status of the freshmen in their cohort. The mentors are encouraged to reflect on their freshmen, see if they feel disoriented, and provide appropriate campus resources or department information. We check in with the mentors and ask them how the instructors can improve the experience in both courses. This mid-semester assessment gives an opportunity to make adjustments and plan additional training for the rest of the semester after collecting suggestions. All the freshmen in the seminar classes likely have received the results of their first exam, so we prepare the mentors to reorient the freshmen by reviewing their SMART goals (Appendix G). Next, we assign the final project groups to the mentors. The first set of log sheets (Appendix L) is due; the purpose is to show the interactions between the mentors and freshmen and help the mentors assess how much they know about their freshmen. From this information, mentors learn about areas they need to adjust.

Freshmen Seminar Class In the freshmen seminar class, the instructor is present, but the mentors facilitate most of the class, covering the topics discussed in the previous weeks. The topics are personal values, academic integrity, and student goals (Appendix G); time management (Appendix H); and campus resources and SMART goals (Appendix G). A brief orientation to the topics is provided, and the mentors facilitate discussion of the topics for the remainder of this class.

Week 7

Mentor Class In the mentor course, we again have the mentors discuss how things are going in facilitating the freshmen's SMART goals (Appendix G) and specific topics. We debrief the mentors on their experiences, discussing what things have gone well and what areas need to be changed in future interactions with their freshmen. During the class, the mentors complete their own individual assessments about learning styles (Appendix M), multiple intelligences (Appendix N), and study skills (Appendix O) based on specific learners' preferences.

Freshmen Seminar Class In the seminar class, there are two scientific talks from department faculty.

Week 8

Mentor Class During this week, context regarding current issues in education is provided (Appendix P). This includes some of the topics discussed in Chap. 1. The mentors have opportunities to learn about common trends associated with socioeconomic status, racial groups, college preparation, diversity, and student success. During this class time, case studies are used to assist the mentors in applying this information to interactions with their freshmen. In addition, time is provided for the mentors to begin preparing for their group projects. The mentors also submit a reflection paper. Both the group project and the individual reflection paper allow time for reflection, or dwell time, when the mentors examine their learning experiences from a different perspective. This dwell time enables them to see the world in different ways.

The mentors begin to address the transition from a large class with multiple cohorts to individual breakout sections in which the freshmen will engage with an individual faculty member for five weeks. During these five weeks, the mentors will meet with their freshmen cohort not during class but in activities outside class and in review sessions. During the mentor class, the mentors conduct their group projects, which are the culmination of the dwell time in which they should be engaging.

Freshmen Seminar Class In the seminar, there are two scientific talks by department faculty.

Week 9

Mentor Class In the mentor class, we ask the mentors how their freshmen are doing and search for any freshmen who are disengaged, disoriented, or not showing up for class. We attempt to connect these freshmen with the appropriate university resources. In this week, a career services staff member gives a presentation on career values assessment for career direction and leadership. The mentors make a list of what is important for a satisfying job or career.

Freshmen Seminar Class In the freshmen seminar class, there are two scientific talks by department faculty. Also, this week, the mentors prepare the freshmen to think about the upcoming breakout groups.

Week 10

Mentor Class The mentors are again queried about the freshman cohort's status to develop group solutions and add extra dwell time or reflection. The mentors are presented with methods for rejuvenating (Appendix Q) and preventing burnout. Just like the freshmen, the mentors have been progressing along their semester, and now with five weeks left (before finals in their other classes), we ask them to dwell and reflect with energy and enthusiasm. These rejuvenation methods are timely for the mentors themselves as well as the freshmen, who have not experienced finals week at a university before. The mentor presentations will start this week or the next. The mentor presentations are group projects in which, with creativity and without overlap, the mentors are required to make a group PowerPoint presentation concerning how their views have changed or remained the same during the 10 weeks of the mentor class.

Freshmen Seminar Class We prepare the freshmen for the breakout groups during the last five weeks at a different time in a different room. In the second half of the class, we ask for feedback, and the mentors provide an initial wrap-up of this stage of the semester. More details are provided in the next section.

Weeks 11–15

Mentor Class In the mentor class, we ask the mentors about their freshmen's status, collect an additional set of log sheets (Appendix L), and continue mentor presentations. The purpose of the presentations is to enable the mentors to work together and to consider all the facets we have taught them, how they have evolved in their 10 weeks with the freshmen, and how they responded to the freshmen. We ask the mentors how their leadership styles have changed. This creates the possibility of sparking new ideas in the mentors' minds. They develop group presentation skills that stretch them beyond their individual intellectual boundaries. We provide dwell time for them to think about leadership and mentorship issues in this new environment of the group presentation. Each group includes four or five students and one or two emeriti mentors. The mentors are encouraged to give a collaborative group presentation and not individual experiences—the presentation should be a group story. The mentors are required to speak for equal amounts of time, and the full presentation should be 25 minutes, followed by five minutes for questions.

As a conclusion to the experience, the mentors have a celebration where they acknowledge the success of their efforts. The event includes PowerPoint slides, certificates, and leadership from faculty and advising staff to recognize the efforts of the mentors and emeriti mentors. In addition, the department director is asked to solicit suggestions from the mentors about what they enjoyed in the experience and suggestions to improve it in future years.

Freshmen Seminar Class During these weeks, an individual faculty member meets weekly with a class of 19 or fewer freshmen in the same cohort group as the first 10 weeks. The philosophy of freshmen–faculty interaction allows for small group interaction needed at institutions. Students often criticize a lack of access to or intimidation regarding attending faculty office hours. This method provides a structured opportunity to interact with faculty and supports retention.

The faculty are assigned five weeks to interact with a maximum of 19 freshmen. They are encouraged to talk about their research and careers. The freshmen are encouraged to ask questions and visit office hours. Faculty are not encouraged to include extensive homework assignments. This is because the freshmen are getting close to finals week, and the homework could provide additional strain on the freshmen. However, the faculty have academic freedom within the five weeks.

Grading

The freshmen seminar learning management platform is set up to indicate the division of points within the 15 weeks. The first 10 weeks can provide 200 points, and the last five weeks are worth 100 points. This proportionally reflects the amount of time the freshmen spend in each portion of the class. The learning management platform records the points collected in categories for the first 10 weeks, and then faculty assign the points for the last five weeks of the class in a separate column. This gives the faculty flexibility to add assignments for the last five weeks if they wish. The total students' points in the first 10 weeks and the last five weeks are used to calculate their overall score.

The first 10 weeks of the semester are largely based on attendance and discussion board posts, in which the students summarize their interpretations of the scientific presentations. This may not appear to be sufficient, but the attendance and the discussion board provides information and dwell time on the content and gives the freshmen information and insights to help them successfully move forward. In addition, the discussion board posts provide dwell time for the freshmen to reflect on their career development and judge whether they are making the planned progress toward their SMART goals (Appendix G). This is the concluding information for the course.

In summary the purpose of the class is to support retention and student success. Although important, the content is secondary to these other lessons. What makes this experience unique and more inviting to the freshmen is the discipline-specific context. Our policy is to be inclusive rather than exclusive and punitive. The value of this course is that it supports retention and helps the freshmen experience success and a sense of belonging.

Throughout this book, we provide information about current issues regarding retention and graduation rates in STEM education. In addition, best practices for the development of a freshmen retention program are shared. The perspectives of the administration, faculty, staff, peer mentors, and freshmen are discussed to include multiple perspectives on designing a retention program. We hope that this information will enable faculty and administrators to make good progress toward improving STEM education.

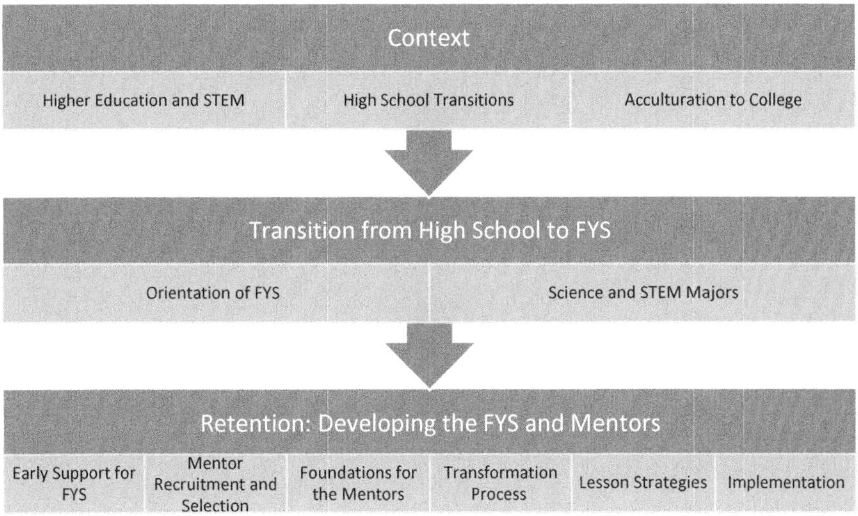

Fig. 11.1 Details of the implementation process for the program summarized in this chapter

Appendices

Appendix A

Ice-breaker Websites

Fun Icebreakers for Meetings
https://www.thebalance.com/fun-ice-breaker-questions-1918413

Top 10 Icebreakers
https://www.thebalance.com/top-ice-breakers-1918426

Icebreakers and Energizers
http://www.training-games.com/pdf/40FreeIceBreakers.pdf

Appendix B

Planning the Session

Name: ______________________________ Date: ______________

Topic: ____________3 SMART Goals: __________ Recitation: ____________

Objectives: What does the group need to accomplish in this session?

Materials to be used: SMART goals, time management

C. Harrell, D. G. Capco, *The STEM Pathway and Student Retention*,
https://doi.org/10.1007/978-3-030-65865-6

Plan: What activities have I planned to use to achieve my objective? Am I incorporating different teaching strategies to reach students with different learning styles?

Icebreaker *(if used)*:

Introduction: *How can I get the students to focus on the topic?*

Equipment needed: *Do I need to arrange for any A/V or computer equipment?*

In-class activities: *Plan at least* ***three****.*

SMART goals:

Closure: *How will I wrap up the topic and bring the class period to a close?*

References used:

Notes: *Follow up after class. Consider what worked and what didn't.*

Appendix C

Facebook Page with Privacy

You have two options:

1) Create a standalone Facebook page. This is quick and easy.

   or

2) Create a Facebook page nested within your institution's page. This has the added benefit of being associated with the institution and is a good jumping-off point for getting involved in extracurricular activities.
   To make a group:

For option 1, follow the directions below.

For option 2, first search for "designated school." Click on the one that is a closed group.

Join this group.
Under the banner on the right is the "create group" icon.
Now follow the directions below.

1. From your home page, go to the **Groups** section on the left side menu and click **Create Group**.

2. A window will appear, where you can add a group name and members and select the privacy settings for your group. Click **Create** when you're done.
3. Once the group is created, you'll be taken to the group page. To get started, click ✱ at the top right of the page and select **Edit Group Settings**. From here, you can add a group description and tags, set up a group email address, post a group picture, and manage members.

Note: You cannot add people who are not your friends, so once you have your group created and ready, send your mentees a link to your page, and they can request to join your group.

From here, you can put as much or as little effort as you want into designing the appearance of your group page. The following is a link to the help page for Facebook groups. It's a good source and shows step-by-step methods to edit your page, add/drop members, poll students, make events, and create documents.

Group help page:
https://www.facebook.com/help/162866443847527/

Privacy

Remember, until the end of the semester, you are your students' mentor, not their friend. It, therefore, is important to maintain your privacy. It is not recommended that you become friends with them on Facebook or allow them full access to your profile. Through your group page, you can maintain consistent communication with your students and post any pictures, posts, or links that you want them to see. The following are recommended privacy settings.

First, here is how you access your privacy settings:

1. Click ⚙ in the upper-right corner of any Facebook page.
2. Select **Privacy Settings** from the dropdown menu.
3. Click on a setting (e.g., **Who can see your future posts?**) to edit it, or use the left column to view your other settings

You can also quickly view and adjust the most widely used privacy settings and tools from your Privacy Shortcuts at the top right of any Facebook page. To control the privacy of posts, photos, and other stuff you share on your timeline, choose your audience when you post. Once in this window, you can change your basic privacy settings. It is up to you whether you want your students to view your posts. Just remember that if you allow them to see your posts, you must keep them professional.

Also in this window, on the left under the privacy icon is a "timelines and tagging" icon. Click on this to determine who can see photos you are tagged in and posts on your timeline. It is highly recommended that you limit this to friends only because you can't regulate what your friends post of you. This is the most likely venue of embarrassment.

For more information, here is a link to Facebook's privacy help page:
https://www.facebook.com/help/325807937506242/

Appendix D

Learning Management Platform Websites

Learning Platforms
https://www.learningplatforms.net/a-z/

Blackboard
http://www.blackboard.com/index.html

Appendix E

Presentation of Leadership Styles

http://changingminds.org/disciplines/leadership/styles/servant_leadership.htm

Appendix F

Risks and Rewards of Mentoring

Rhodes, J. E. (2002). *Stand by me: The risks and rewards of mentoring today's youth*. Cambridge, MA: Harvard University Press.

Appendix G

SMART Goals

The quality of the goals we set determine their usefulness. According to the SMART System, developed by Kenneth Blanchard and Spencer Johnson, the most effective, useful goals meet five key criteria included in the resources.

A Motivational Story with Wisdom: Dreams, Unknown Author
http://www.wow4u.com/dreams12/index.html

Kenneth Blanchard's Goal-Setting Resources
https://www.kenblanchardbooks.com/

Appendix H

Time-management Resources

Time-management Calendars
https://www.bing.com/images/search?q=time+management+calendar&qpvt=Time+management+calendar&FORM=IGRE

Worksheets
http://www.pacific.edu/Documents/student-life/Student%20Support%20Services/TimeManagement%20Worksheet.pdf

Appendix I

Tips for Facilitating an Effective Discussion

http://www.americanheart.org/downloadable/heart/1103321531845Facilitating_Effective_Discussion_GWTG.pdf

Appendix J

Letter to a Young Scientist

Schatz, G. (2004). "Letter to a young scientist" by Gottfried. FEBS Letters, Vol. 558. doi:10.1016/S0014-5793(03)01500-X

Appendix K

Dwell Time on Leadership

Heider, J. (1985). *The Tao of leadership*. New York, NY: Humanics Limited.
Heider, J. (2015). *The Tao of leadership*. Palm Beach, FL: Green Dragon Books.

Relevant chapter readings from the *Tao of Leadership* (1985):
2, 3, 5, 8, 9, 10, 11, 12, 13, 16, 17, 23, 24, 64, and 81

Appendix L

Log Sheet

					Date: ____________
	Student Name		**Date of Contact**	**Total Hours**	**Description of the Interaction**
#	Last Name	First Name	(Month/Day/ Year):	#	What happened? Where did it happen? Who was there? What did you learn about the student? What did you learn about yourself?
1	**Smith**	**Jim**	**10/31/2012**	**2**	
2					
3					
4					
5					
6					
7					
8					

Appendix M

Leadership Styles

http://www.engr.ncsu.edu/learningstyles/ilsweb.html

Appendix N

Multiple Intelligences

Multiple Intelligences
http://www.tecweb.org/styles/gardner.html
http://multipleintelligencesoasis.org/

Multiple Intelligences Self-Assessment
https://www.edutopia.org/multiple-intelligences-assessment
http://www.literacynet.org/mi/assessment/findyourstrengths.html

Appendix O

Study Skills

TED Talk: The First 20 Hours
https://ed.ted.com/on/Imq14bHp

Study Guides and Strategies
http://www.studygs.net/

Appendix P

Issues in Education

Handelsman, J., Pfund, C., Lauffer, S., & Pribbenow, C. (2005). Teaching scientists to teach. In H. Handelsman (Ed.), *Entering mentoring: A seminar to train a new generation of scientists*. Madison, WI: Board of Regents of the University of Wisconsin System.

Social Class and the Hidden Curriculum of Work
http://www1.udel.edu/educ/whitson/897s05/files/hiddencurriculum.htm

Appendix Q

Rejuvenation

Stress Relief Video
http://www.youtube.com/user/LivestrongWoman?v=IdR9bFBEx4g

Tim Tialdo's Positive Self-talk
https://www.youtube.com/watch?v=zXuR2WUH4Fc

References

Abdul-alim, J. (2011). Mentor Program Introduces Students to Stem-Related Fields; students exposed to careers in architecture, engineering. *Education Week, 30*(17), 1.

Abrams, E., Taylor, P. C., & Guo, C.-J. (2014). Contextualizing culturally relevant science and mathematics teaching for indigenous learning. *International Journal of Science and Mathematics Education, 11*. https://doi.org/10.1007/s10763-012-9388-2.

Allen, D., & Tanner, K. (*2005*). Approaches to biology teaching and learning: From a scholarly approach to teaching to the scholarship of teaching. *Cell Biology Education, 4*, 1–6.

Alonso, M. A., Castaño, G., Calles, A. M., & Sánchez-Herrero, S. (2010). Assessment of the efficacy of a peer mentoring program in a university setting. *Spanish Journal of Psychology, 13*(2), 685–696.

Andrade, E., McKeever, A., Rivera, R., Withers, E., & Woo, H. (2020). Millennials and Moral Panic in the United States and beyond. *Sociological Perspectives, 63*(3), 496–505. https://doi.org/10.1177/0731121420919590.

Arizona Minority Education Policy Analysis Center. (2007). *Minority student progress report: A snapshot of Arizona's educational achievement*. Phoenix, AZ: AMEPAC.

Armbruster, P., Patel, M., Johnson, E., & Weiss, M. (2009). Active learning and student-centered pedagogy improve student attitudes and performance in introductory biology. *CBE Life Sciences Education, 8*(3), 203–213. https://doi.org/10.1187/cbe.09-03-0025.

Arum, R., & Roksa, J. (2011). *Academically adrift: Limited learning on college campuses*. University of Chicago Press.

Astin, A. W. (1975). *Preventing students from dropping out*. San Francisco, CA: Jossey-Bass.

Astin, A. W. (1984). Student involvement: A developmental theory for higher education. *Journal of College Student Personnel, 25*, 297–308.

Astin, A. W. (1993). *What matters in college? Four critical years revisited*. San Francisco, CA: Jossey-Bass.

Atkins, K., Dougan, B., Dromgold-Sermen, M., Potter, H., Sathy, V., & Panter, A. (2020). "Looking at Myself in the Future": how mentoring shapes scientific identity for STEM students from underrepresented groups. *International Journal of STEM Education, 7*(1), 1–15. https://doi.org/10.1186/s40594-020-00242-3.

Aveling, N. (2006). 'Hacking at our very roots': rearticulating White racial identity within the context of teacher education. *Race Ethnicity and Education, 9*(3), 261–274. https://doi.org/10.1080/13613320600807576.

Bandura, A. (1997). *Self-efficacy: The exercise of control*. New York: W.H. Freeman.

Banks, C. A. (1995). Intellectual leadership and the influence of early African American scholars on multicultural education. *Educational Policy, 9*(3), 260–280. https://doi.org/10.1177/0895904895009003003.

C. Harrell, D. G. Capco, *The STEM Pathway and Student Retention*,
https://doi.org/10.1007/978-3-030-65865-6

Barr, R., & Tagg, J. (1995). From Teaching to Learning — A New Paradigm For Undergraduate Education. *Change: The Magazine of Higher Learning, 27*(6), 12–26. https://doi.org/10.1080/00091383.1995.10544672.

Beltman, S., & Schaeben, M. (2012). Institution-wide peer mentoring: Benefits for mentors. *The International Journal of the First Year in Higher Education, 3*(2), 33–44. https://doi.org/10.5204/intjfyhe.v3i2.124.

Bhatt, R., West, B., & Chaudhary, S. (2020). Biomedical career enrichment programs: Exploring women and minority participants' motivators and outcomes. *PLoS One, 15*(2), e0228934. https://doi.org/10.1371/journal.pone.0228934.

Black, A., Terry, N., & Buhler, T. (2016). The impact of specialized courses on student retention as part of the freshman experience. *Academy of Educational Leadership Journal, 20*(1), 85.

Boaler, J. (2002). Learning from teaching: Exploring the relationship between reform curriculum and equity. *Journal for Research in Mathematics Education, 33*(4), 239–258. https://doi.org/10.2307/749740.

Bogue, G., & Aper, J. (2000). *Exploring the heritage of American higher education: The evolution of philosophy and policy*. Phoenix, AZ: Oryx Press.

Borda, E., Schumacher, E., Hanley, D., Geary, E., Warren, S., Ipsen, C., & Stredicke, L. (2020). Initial implementation of active learning strategies in large, lecture STEM courses: lessons learned from a multi-institutional, interdisciplinary STEM faculty development program. *International Journal of STEM Education, 7*(1), 1–18. https://doi.org/10.1186/s40594-020-0203-2.

Boud, D., & Feletti, G. (1991). *The challenge of problem-based learning*. New York: St. Martin's Press.

Boud, D. J., & Higher Education Research and Development Society of Australasia. (1985). *Problem-based learning in education for the professions*. Sydney: HERDSA.

Bowden, M. P., Abhayawansa, S., & Bahtsevanoglou, J. (2015). Overconfidence of vocational education students when entering higher education. *Education and Training, 57*(4), 429–447. https://doi.org/10.1108/ET-02-2014-0012.

Bowen, B. (1999). A report card on diversity: Lessons for business from higher education. *Harvard Business Review, 77*(1), 138–149.

Bowman, L., Culhane, K., Park, A., & Kucera, K. (2019). Course-based prefaculty training program introduces instructional methods, increases instructor self-efficacy, and promotes professional development. *Biochemistry and Molecular Biology Education, 48*(2), 156–167. https://doi.org/10.1002/bmb.21326.

Brandsford, J. D., Pellegrino, J. W., Donovan, S., & National Research Council (U.S.). (1999). *How people learn: Bridging research and practice*. Washington, DC: National Academy Press.

Braxton, J. (2019). Leaving college: Rethinking the causes and cures of student attrition by Vincent Tinto (review). *Journal of College Student Development, 60*(1), 129–134. https://doi.org/10.1353/csd.2019.0012.

Brint, S., Cantwell, A. M., & Saxena, P. (2012). Disciplinary categories, majors, and undergraduate academic experiences: Rethinking Bok's "underachieving colleges" thesis. *Research in Higher Education, 53*(1), 1–25. https://doi.org/10.1007/s11162-011-9227-2.

Brown, S. K., & Hirschman, C. (2006). The end of affirmative action in Washington state and its impact on the transition from high school to college. *Sociology of Education, 79*(2), 106–130. https://doi.org/10.1177/003804070607900202.

Byrne, E. M. (1978). *Women and education*. London: Tavistock Publications.

Carlson, S. M. (2005). Developmentally sensitive measures of executive function in preschool children. *Developmental Neuropsychology, 28*(2), 595–616. https://doi.org/10.1207/s15326942dn2802_3.

Carter, R. T., & Goodwin, A. L. (1994). Racial identity and education. *Review of Research in Education, 20(1)*, 291–336. https://doi.org/10.2307/1167387.

Castellanos, J., & Jones, L. (2003). *The majority in the minority*. Sterling, VA: Stylus.

Charleston, L., George, P., Jackson, J., Berhanu, J., & Amechi, M. (2014). Navigating underrepresented STEM spaces: Experiences of black women in US computing science higher education

programs who actualize success. *Journal of Diversity in Higher Education, 7*(3), 166–176. https://doi.org/10.1037/a0036632.
Chelberg, K., & Bosman, L. (2020). American Indian college student mentoring: A study to measure changes in self-efficacy. *Educational Sciences, 10*(2), 38. https://doi.org/10.3390/educsci10020038.
Chen, Y., & Upah, S. (2020). Data analytics and STEM student success: The impact of predictive analytics-informed academic advising among undeclared first-year engineering students. *Journal of College Student Retention: Research, Theory & Practice, 22*(3), 497. https://doi.org/10.1177/1521025118772307.
Chester, A., Burton, L. J., Xenos, S., & Elgar, K. (2013). Peer mentoring: Supporting successful transition for first year undergraduate psychology students. *Australian Journal of Psychology, 65*(1), 30–37. https://doi-org.ezproxy1.lib.asu.edu/10.1111/ajpy.12006.
Chitiga, M., Kaniuka, T., & Ombonga, M. (2019). How do millennials learn?: Implications for higher education pedagogy. *International Journal of Information and Communication Technology Education (IJICTE), 15*(1), 29–41. https://doi.org/10.4018/IJICTE.2019010103.
Christe, B. (2013). The importance of faculty-student connections in STEM disciplines: A literature review. *Journal of STEM Education: Innovations and Research, 14*(3), 22.
Cintrón, R., Oosahwe, E. S. L., & Shotton, H. J. (2007). Stories of success: Experiences of American Indian students in a peer-mentoring retention program. *The Review of Higher Education, 31*(1), 81–107. https://doi.org/10.1353/rhe.2007.0060.
College Board. (2011). The 7th annual AP report to the nation. Retrieved from http://apreport.collegeboard.org/sites/default/files/downloads/pdfs/AP_RTN_2011.pdf
Colomer, J., Serra, T., Cañabate, D., & Bubnys, R. (2020). Reflective learning in higher education: Active methodologies for transformative practices. *Sustainability (Basel, Switzerland), 12*(9), 3827. https://doi.org/10.3390/su12093827.
Colvin, J. W., & Ashman, M. (2010). Roles, risks, and benefits of peer mentoring relationships in higher education. *Mentoring and Tutoring: Partnership in Learning, 18*(2), 121–134. https://doi.org/10.1080/13611261003678879.
Corrêa, B. B., Pinto, P. R., & Rendas, A. B. (2003). How do learning issues relate with content in a problem-based learning pathophysiology course? *Advances in Physiology Education, 27*(2), 62–69. https://doi.org/10.1152/advan.00021.2002.
Council, L., Johnson, C., Santellano, K., & Yazdiha, H. (2020). Linking contexts, intersectionality, and generations: Toward a multidimensional theory of millennials and social change. *Sociological Perspectives, 63*(3), 486–495. https://doi.org/10.1177/0731121420914294.
Cutright, T. J., Evans, E., & Brantner, J. S. (2014). Building an undergraduate STEM team using team-based learning leading to the production of a storyboard appropriate for elementary students. *Journal of Science Education and Technology, 23*(3), 344–354. https://doi.org/10.1007/s10956-013-9467-3.
Denial, A., Schreiner, L., Kammer, R., & Kim, Y. K. (2015). The validation of the active learning in health professions scale. *Interdisciplinary Journal of Problem-Based Learning, 9*(1), 58–72. https://doi.org/10.7771/1541-5015.1504.
Dewey, J. (1916). *Democracy and education: An introduction to the philosophy of education*. New York: Macmillan.
Drake, T. M., & Osborne, N. L. (2007). *Minority student success report 2007*. Retrieved from http://www.azhighered.org/Content/AMEPAC%20Report%20PDF%20for%20Web%20Site%2010-22-07%20FINAL.pdf
Elliott, C., Mavriplis, C., & Anis, H. (2020). An entrepreneurship education and peer mentoring program for women in STEM: mentors' experiences and perceptions of entrepreneurial self-efficacy and intent. *International Entrepreneurship and Management Journal, 16*(1), 43–67. https://doi.org/10.1007/s11365-019-00624-2.
Ensher, E. A., & Murphy, S. E. (2011). The mentoring relationship challenges scale: The impact of mentoring stage, type, and gender. *Journal of Vocational Behavior, 79*(1), 253–266.

Falloon, G., Hatzigianni, M., Bower, M., Forbes, A., & Stevenson, M. (2020). Understanding K 12 STEM education: A framework for developing STEM literacy. *Journal of Science Education and Technology, 29*(3), 369–385. https://doi.org/10.1007/s10956-020-09823-x.

Frehill, L. M. (1997). Education and occupational sex segregation: The decision to major in engineering.

The Sociological Quarterly, 38(2), 225–249. Retrieved from http://www.jstor.org.ezproxy1.lib.asu.edu/stable/4120734.

Friedrich, J. M. (2014). A classroom-based distributed workflow initiative for the early involvement of undergraduate students in scientific research. *Journal of Science Education and Technology, 23*(1), 59–66. https://doi.org/10.1007/s10956-013-9450-z.

Fries-Britt, S., & White-Lewis, D. (2020). In pursuit of meaningful relationships: How Black males perceive faculty interactions in STEM. *The Urban Review, 52*(3), 521–540. https://doi.org/10.1007/s11256-020-00559-x.

Fuentes, M., Alvarado, A., Berdan, J. L., & Deangelo, L. (2014). Mentorship matters: Does early faculty contact lead to quality faculty interaction? *Research in Higher Education, 55*. https://doi.org/10.1007/s11162-013-9307-6.

Fromme, K., Corbin, W. R., & Kruse, M. I. (2008). Behavioral risks during the transition from high school to college. *Developmental Psychology, 44*(5), 1497–1504. http://dx.doi.org.ezproxy1.lib.asu.edu/10.1037/a0012614.

Gaddis, S. M. (2012). What's in a relationship? An examination of social capital, race and class in mentoring relationships. *Social Forces, 90*(4), 1237–1269. https://doi.org/10.1093/sf/sos003.

Gahagan, J., & Hunter, M. S. (2006). The second-year experience: Turning attention to the academy's middle children. *About Campus, 11*(3), 17–22. https://doi.org/10.1002/abc.168.

Gandara, P., Horn, C., & Orfield, G. (2005). The access crisis in higher education. *Educational Policy, 19*(2), 255–261. https://doi.org/10.1177/0895904804274060.

Gardner, J. N., & Association of Governing Boards of Universities and Colleges, Washington, DC (2001). *Focusing on the first-year student.* Distributed by ERIC Clearinghouse.

Ghosh-Dastidar, U., & Liou-Mark, J. (2014). Bridging pathways through research and leadership for underrepresented students in STEM. *Mathematics and Computer Education, 48*(3), 214.

Girves, J. E., Zepeda, Y., & Gwathmey, J. K. (2005). Mentoring in a post-affirmative action world. *Journal of Social Issues, 61*(3), 449–479. https://doi.org/10.1111/j.1540-4560.2005.00416.x.

Glasser, C., Merrow, J., & Home Video, P. B. S. (2005). *Declining by degrees: Higher education at risk.* United States: PBS Home Video.

Gonzalez, H. B., Kuenzi, J. J., Congressional Research Service, & Library of Congress. (2012). *Science, technology, engineering, and mathematics (STEM) education: A primer.* Washington, DC: Congressional Research Service.

Grabinger, S., Dunlap, J. C., & Duffield, J. A. (2011, 1997). Rich environments for active learning in action: Problem-based learning. *Research in Learning Technology, 5*(2). https://doi.org/10.3402/rlt.v5i2.10558.

Graham, M. J., Frederick, J., Byars-Winston, A., Hunter, A.-B., & Handelsman, J. (2013). Increasing persistence of college students in stem (2013). Retrieved from http://www.fgcu.edu/STEM/files/Increasing_Persistence_of_College_Students_in_ST.pdf

Gratz v. Bollinger. (2003). 539 U.S. Certiorari to the United States Court of Appeals for the Sixth Circuit 539.

Graunke, S. S., & Woosley, S. A. (2005). An exploration of the factors that affect the academic success of college sophomores. *College Student Journal, 29*, 367–376.

Gregg-Jolly, L., Swartz, J., Iverson, E., Stern, J., Brown, N., & Lopatto, D. (2016). Situating second-year success: Understanding second-year STEM experiences at a liberal arts college. *CBE Life Sciences Education, 15*(3), ar43. https://doi.org/10.1187/cbe.16-01-0044.

Griffith, A. L. (2010). Persistence of women and minorities in STEM field majors: Is it the school that matters? *Economics of Education Review, 29*(6), 911–922.

Grutter v. Bollinger. (2003). 539 U.S. Certiorari to the United States Court of Appeals for the Sixth Circuit 306.

Gunn, L., ter Horst, E., Markossian, T., & Molina, G. (2020). Associations between majors of graduating seniors and average SATs of incoming students within higher education in the U.S. *Heliyon, 6*(5), e03956. https://doi.org/10.1016/j.heliyon.2020.e03956.

Gureasko-Moore, S., DuPaul, G. J., & White, G. P. (2006). The effects of self-management in general education classrooms on the organizational skills of adolescents with ADHD. *Behavior Modification, 30*(2), 159–183. https://doi.org/10.1177/0145445503259387.

Haberman, M. (1991). The pedagogy of poverty versus good teaching. *The Phi Delta Kappan, 73*(4), 290–294. Retrieved January 25, 2021, from http://www.jstor.org/stable/20404620.

Hale, S. (2011). The positive side of community interpreting: An Australian case study. *Interpreting International Journal of Research and Practice in Interpreting, 13*(2), 234–248. https://doi.org/10.1075/intp.13.2.04hal.

Hall, R., & Jaugietis, Z. (2011). Developing peer mentoring through evaluation. *Innovative Higher Education, 36*(1), 41–52. https://doi.org/10.1007/s10755-010-9156-6.

Han, J., & Neuharth-Pritchett, S. (2010). Beliefs about classroom practices and Teachers' education level: An examination of developmentally appropriate and inappropriate beliefs in early childhood classrooms. *Journal of Early Childhood Teacher Education, 31*(4), 307–321. https://doi.org/10.1080/10901027.2010.523775.

Han, H. S., West-Olatunji, C., & Thomas, M. S. (2011). Use of racial identity development theory to explore cultural competence among early childhood educators. *SRATE Journal, 20*(1), 1.

Handelsman, J., Pfund, C., Lauffer, S., & Pribbenow, C. (2005). In H. Handelsman (Ed.), *Entering mentoring: A seminar to train a new generation of scientists*. Madison, WI: Board of Regents of the University of Wisconsin System.

Harper, S. R., Carini, R. M., Bridges, B. K., & Hayek, J. C. (2004). Gender differences in student engagement among African American undergraduates at historically Black colleges and universities. *Journal of College Student Development, 45*(3), 271–284. Johns Hopkins University Press. Retrieved from Project MUSE database.

Harris, F., Palmer, R. T., & Struve, L. (2011). "Cool posing" on campus: A qualitative study of masculinities and gender expression among black men at a private research institution. *Journal of Negro Education, 80*, 47–62.

Helms, J. E. (1990). *Black and White racial identity: Theory, research, and practice*. New York: Greenwood Press.

Helms, J. E. (2008). Chapter 7: A model of white identity development. In *A race is a Nice thing to have* (pp. 29–39). Alexandria, VA: Microtraining Associates. Retrieved from Counseling and Therapy in Video, Volume IV database.

Herrera, F. A., & Hurtado, S. (2011). *Maintaining initial interests: Developing science, technology, engineering, and mathematics (STEM) career aspirations among underrepresented racial minority students*. New Orleans, LA: Association for Educational Research annual meeting.

Hilton, T. L., & Lee, V. E. (1988). Student interest and persistence in science: Changes in the educational pipeline in the last decade. *The Journal of Higher Education, 59*(5), 510–526. Web.

Hodgetts, K. (2008). Underperformance or 'Getting it Right'? Constructions of gender and achievement in the Australian inquiry into Boys' education. *British Journal of Sociology of Education, 29*(5), 465–477. Retrieved January 25, 2021, from http://www.jstor.org/stable/40375370.

Holland, J. M., Major, D. A., & Orvis, K. A. (2012). Understanding how peer mentoring and capitalization link STEM students to their majors. *Career Development Quarterly, 60*(4), 343–354. https://doi.org/10.1002/j.2161-0045.2012.00026.x.

Hunter, A., Laursen, S. L., & Seymour, E. (2007). Becoming a scientist: The role of undergraduate research in students' cognitive, personal, and professional development. *Science Education, 91*, 36–74.

Hutchinson, K. H. (2003). The teach act: Copyright law and online education. *New York University Law Review, 78*, 2204–2305.

Ishitani, T. T. (2006). Studying attrition and degree completion behavior among first-generation college students in the United States. *The Journal of Higher Education, 77*(5), 861–885. https://doi.org/10.1080/00221546.2006.11778947.

Iverson, E., Bragg, D., & Eddy, P. (2020). How faculty change agents enact midlevel leadership in STEM. *New Directions for Community Colleges, 2020*(191), 67–79. https://doi.org/10.1002/cc.20407.

Jacob, S. A., Khan, T. M., Pusparajah, P., Velu, S. S., Lee, L. H., & Davey, T. M. (2016). Students' perceived predictors of an effective active learning/problem-based learning session: A pilot study. *Journal of Pharmacy Practice and Research, 46*(1), 42–46. https://doi.org/10.1002/jppr.1150.

Janke, K., Nelson, M., Bzowyckyj, A., Fuentes, D., Rosenberg, E., & DiCenzo, R. (2016). Deliberate integration of student leadership development in doctor of pharmacy programs. *American Journal of Pharmaceutical Education, 80*(1), 1. https://doi.org/10.5688/ajpe8012.

Jansen, E., André, S., & Suhre, C. (2013). Readiness and expectations questionnaire: A cross-cultural measurement instrument for first-year university students. *Educational Assessment, Evaluation and Accountability, 25*(2), 115–130. https://doi.org/10.1007/s11092-013-9161-2.

Jasper, M. (1996). The first year as a staff nurse: The experiences of a first cohort of project 2000 nurses in a demonstration district. *Journal of Advanced Nursing, 24*(4), 779–790. https://doi.org/10.1046/j.1365-2648.1996.25517.x.

Jenkins, D. M. (2013). Exploring instructional strategies in student leadership development programming. *Journal of Leadership Studies, 6*(4), 48–62. https://doi.org/10.1002/jls.21266.

Jerrim, J. (2014). The unrealistic educational expectations of high school pupils: Is America exceptional? *The Sociological Quarterly, 55*(1), 196–231. https://doi.org/10.1111/tsq.12049.

Johansson, C. (2014). *Transforming students: Fulfilling the promise of higher education*. Baltimore, MD: Johns Hopkins University Press.

Jones, L., Castellanos, J., & Cole, D. (2002). Examining the ethnic minority student experience at predominantly white institutions: A case study. *Journal of Hispanic Higher Education, 1*(1), 19–39. https://doi.org/10.1177/1538192702001001003.

Judge, T. A., & Hurst, C. (2007). Capitalizing on one's advantages: Role of core self-evaluations. *Journal of Applied Psychology, 92*(5), 1212–1227. https://doi.org/10.1037/0021-9010.92.5.1212.

Jungblut, J., Vukasovic, M., & Steinhardt, I. (2018). Higher education policy dynamics in turbulent times – access to higher education for refugees in Europe. *Studies in Higher Education (Dorchester-on-Thames), 45*(2), 327–338. https://doi.org/10.1080/03075079.2018.1525697.

Kalevitch, M., Maurer, C., Badger, P., Holdan, G., & Sirinterlikci, A. (2015). Building a community of scholars: One university's story of students engaged in learning science, mathematics, and engineering through a NSF S-STEM grant – Part II. *Journal of STEM Education: Innovations and Research, 16*(2), 40.

Kelly v. Arriba Soft Corp. (n.d.) No. 00-55521 (United States Court of Appeals for the Ninth Circuit).

Kelly, M. A., Hopwood, N., Rooney, D., & Boud, D. (2016). Enhancing students' learning through simulation: Dealing with diverse, large cohorts. *Clinical Simulation in Nursing, 12*(5), 171–176. https://doi.org/10.1016/j.ecns.2016.01.010.

Kendricks, K., & Arment, A. (2011). Adopting a K-12 family model with undergraduate research to enhance STEM persistence and achievement in underrepresented minority students. *Journal of College Science Teaching, 41*(2), 22–27.

Kendricks, K. D., Nedunuri, K. V., & Arment, A. R. (2013). Minority student perceptions of the impact of mentoring to enhance academic performance in STEM disciplines. *Journal of STEM Education: Innovations and Research, 14*(2), 38.

Kenedy, R., Monty, V., & Lambert-Drache, M. (2012). Transitions through pastoral peer mentoring: A qualitative analysis of the challenges and successes. *Journal of the First-Year Experience & Students in Transition, 24*(1), 79.

Keneley, M., & Jackling, B. (2011). The acquisition of generic skills of culturally-diverse student cohorts. *Accounting Education, 20*(6), 605–623. https://doi.org/10.1080/09639284.2011.611344.

Kent, M., & Leaver, T. (2014). *An education in Facebook?: Higher education and the world's largest social network*. New York: Routledge.

Kersh, M. E., & Masztal, N. B. (1998). An analysis of studies of collaboration between universities and K-12 schools. *The Educational Forum, 62*(3), 218.

Keup, J. R. (2005). The impact of curricular interventions on intended second year re-enrollment. *Journal of College Student Retention: Research, Theory & Practice, 7*(1), 61–89. https://doi.org/10.2190/2DCU-KABY-WVQH-2F8J.

Keup, J. R., & Stolezenberg, E. B. (2004). The 2003 Your First College Year (YFCY) survey: Exploring the academic and personal experiences of first-year students (Monograph No. 40). Columbia, SC: University of South Carolina, National Resource Center for the First-Year Experience and Students in Transition.

Kezar, A. J. (2011). *Recognizing and serving low-income students in higher education: An examination of institutional policies, practices, and culture*. New York: Routledge.

Khatri, D. S., Hughes, A. (2013). *A teaching guide to revitalizing STEM education: Phoenix in the classroom*. Lanham, MD: Rowman & Littlefield Education.

Khoiriyah, U., Roberts, C., Jorm, C., & Vleuten, C. P. v. d. (2015). Enhancing students' learning in problem based learning: Validation of a self-assessment scale for active learning and critical thinking. *BMC Medical Education, 15*(1), 140. https://doi.org/10.1186/s12909-015-0422-2.

Kienzl, G., & Trent, W. (2009). *Underrepresented undergraduates in STEM: From matriculation to degree completion at large, research-intensive public universities*. Report to the Alfred P. Sloan Foundation, New York, NY.

Kim, S. (2007). Learning goal orientation, formal mentoring, and leadership competence in HRD. *Journal of European Industrial Training, 31*(3), 181.

Kinner, D., & Lord, M. (2018). Student-perceived gains in collaborative, course-based undergraduate research experiences in the geosciences. *Journal of College Science Teaching, 48*(2), 48–58.

Kiouses, D. (2014). *Dual credit: Easing the transition from high school to college*. ProQuest Dissertations Publishing.

Kiyama, J. M., Luca, S. G., Raucci, M., & Crump-Owens, S. (2014). A cycle of retention: Peer mentors' accounts of active engagement and agency. *College Student Affairs Journal, 32*(1), 81.

Kling, T. P., & Salomone, M. (2015). Creating a peer-led cooperative learning program to improve STEM retention. *Change: The Magazine of Higher Learning, 47*(6), 42. https://doi.org/10.1080/00091383.2015.1089758.

Koch, R. A., & Loverude, M. (2015). STEM student success through the CSUF catalyst center. *Peer Review, 17*(2), 25.

Koenig, K., Schen, M., Edwards, M., & Bao, L. (2012). Addressing STEM retention through a scientific thought and methods course. *Journal of College Science Teaching, 41*(4), 23.

Kokkelenberg, E. C., & Sinha, E. (2010). Who succeeds in STEM studies? An analysis of Binghamton University undergraduate students. *Economics of Education Review, 29*(6), 935–946. https://doi.org/10.1016/j.econedurev.2010.06.016.

Kolvoord, B., Phillips, K., Puffenbarger, R., Ryan, L., McGhee, R., Sowers, J., et al. (2016). Bridging the valley: Recruiting and retaining STEM majors. *Journal of STEM Education: Innovations and Research, 17*(4), 8.

Komives, S. R. (2011). *The handbook for student leadership development* (2nd ed.). San Francisco: Jossey-Bass.

Konkel, L. (2015). Racial and ethnic disparities in research studies the challenge of creating more diverse cohorts. *Environmental Health Perspectives, 123*(12), A297–A302. https://doi.org/10.1289/ehp.123-A297.

Koo, S., & Nishimura, T. S. (2013). Minority within a minority paradox: Asian experiences in Latino schools & communities. *Multicultural Education, 20*(2), 17–26.

Kotamraju, P., & Blackman, O. (2011). Meeting the 2020 American Graduation Initiative (AGI) Goal of increasing postsecondary graduation rates and completions: A macro perspective of community college student educational attainment. *Community College Journal of Research and Practice, 35*(3), 202–219.

Kötter, T., Tautphäus, Y., Obst, K. U., Voltmer, E., & Scherer, M. (2016). Health-promoting factors in the freshman year of medical school: A longitudinal study. *Medical Education, 50*(6), 646–656. https://doi.org/10.1111/medu.12987.

Kraal, D. (2016). Legal teaching methods to diverse student cohorts: A comparison between the United Kingdom, The United States, Australia and New Zealand. *Cambridge Journal of Education*, 1–23. https://doi.org/10.1080/0305764X.2016.1190315.

Kricorian, K., Seu, M., Lopez, D., Ureta, E., & Equils, O. (2020). Factors influencing participation of underrepresented students in STEM fields: matched mentors and mindsets. *International Journal of Stem Education, 7*(1), 1–9. https://doi.org/10.1186/s40594-020-00219-2.

Krieg, D. B. (2013). High expectations for higher education? Perceptions of college and experiences of stress prior to and through the college career. *College Student Journal, 47*(4), 635–643.

Kristjánsson, K. (2007). Justified self-esteem. *Journal of Philosophy of Education, 41*(2), 247–261.

Kuh, G. D., & Documenting Effective Educational Practice (Project). (2005). *Student success in college: Creating conditions that matter* (1st ed.). San Francisco, CA: Jossey-Bass.

Kuh, G. D. (2010). *Student success in college: Creating conditions that matter* (1st ed.). Chichester; San Francisco, CA: Jossey-Bass.

Kuh, G. D., Schneider, C. G., & Association of American Colleges and Universities. (2008). *High-impact educational practices: What they are, who has access to them, and why they matter.* Washington, DC: Association of American Colleges and Universities.

Lacey, T. A., & Wright, B. (2009). Employment outlook: 2008-18-occupational employment projections to 2018. *Monthly Labor Review, 132*(11), 82–123. Retrieved from https://heinonline.org/HOL/P?h=hein.journals/month132&i=1377.

Lake, P. F. (1999). The rise of duty and the fall of in loco parentis and other protective tort doctrines in higher education law. *Missouri Law Review, 64*, 1. Retrieved from https://heinonline.org/HOL/P?h=hein.journals/molr64&i=11.

Lang, K. B. (2012). The similarities and differences between working and non-working students at a mid-sized American public university. *College Student Journal, 46*(2), 243.

Lang, M., Ford, C. A., & Florida Agricultural and Mechanical University. (1988). *Black student retention in higher education.* Springfield, IL: C.C. Thomas.

Larose, S., Cyrenne, D., Garceau, O., Harvey, M., Guay, F., Godin, F., et al. (2011). Academic mentoring and dropout prevention for students in math, science and technology. *Mentoring and Tutoring: Partnership in Learning, 19*(4), 419–439. https://doi.org/10.1080/13611267.2011.622078.

Laurs, D., Eggeling, D., & Harris, P. (2013). Empowering student leaders to nurture the first year experience through cross-cultural diversity. A practice report nau te raurau, naku te raurau, ka ora ai te iwi. *The International Journal of the First Year in Higher Education, 4*(2), 105–110. https://doi.org/10.5204/intjfyhe.v4i2.176.

Law, H., Ireland, S., & Hussain, Z. (2007, 2013). *The psychology of coaching, mentoring and learning* (2nd). Hoboken, NJ; Chichester, UK: John Wiley & Sons Ltd.

Lawrence, S. M., & Tatum, B. D. (1997). Teachers in transition: The impact of antiracist professional development on classroom practice. *Teachers College Record, 99*(1), 162–178. Retrieved from https://eric.ed.gov/?id=EJ557693.

Layton, L. (2014). U.S. graduation rate hits high. *The Washington Post,* A.3.

Leary, M., Morewood, A., & Bryner, R. (2020). A controlled intervention to improve freshman retention in a STEM-based physiology major. *Advances in Physiology Education, 44*(3), 334–343. https://doi.org/10.1152/advan.00038.2020.

Leidenfrost, B., Strassnig, B., Schabmann, A., Spiel, C., & Carbon, C. (2011). Peer mentoring styles and their contribution to academic success among mentees: A person-oriented study in higher education. *Mentoring and Tutoring: Partnership in Learning, 19*(3), 347–364. https://doi.org/10.1080/13611267.2011.597122.

Leidenfrost, B., Strassnig, B., Schütz, M., Carbon, C., & Schabmann, A. (2014). The impact of peer mentoring on mentee academic performance: Is any mentoring style better than no mentoring at all? *International Journal of Teaching and Learning in Higher Education, 26*(1), 102.

Lent, R. W., Miller, M. J., Smith, P. E., Watford, B. A., Lim, R. H., & Hui, K. (2016). Social cognitive predictors of academic persistence and performance in engineering: Applicability across

gender and race/ethnicity. *Journal of Vocational Behavior, 94*, 79–88. https://doi.org/10.1016/j.jvb.2016.02.012.
Leon, A. (2011). *Collegiate student athlete success in a first year experience learning community* (Order No. 3494410). Available from ProQuest Dissertations & Theses Global. (920878919).
Liljequist, L., & Stone, S. (2009). Measuring the success of a summer reading program: A five-year study. *Journal of the First-Year Experience & Students in Transition, 21*(2), 87.
Liou-Mark, J., Dreyfuss, A. E., Blake, R. A., Lansiquot, R. D., & Yu, K. (2013). Navigation by mentoring and leadership: Sustaining women majoring in mathematics. *Mathematics and Computer Education, 47*(2), 134.
Lipinski, T. A. (2003). The climate of distance education in the 21st century: Understanding and surviving the changes brought by the TEACH (technology, education, and copyright harmonization) act of 2002. *The Journal of Academic Librarianship, 29*(6), 362–374. https://doi.org/10.1016/j.jal.2003.08.003.
Liu, L., Wagner, W., Sonnenberg, B., Wu, X., & Trautwein, U. (2014). Independent freshman admission and educational inequality in the access to elite higher education. *Chinese Sociological Review, 46*(4), 41–67. https://doi.org/10.2753/CSA2162-0555460403.
Lochbaum, A. (2010). *Self-discrepancies, symbolic self-completion, and the role of possessions in the transition from high school to college*. (Order No. 1488186). Available from ProQuest Dissertations & Theses Global. (849735595) .
Loots, A. (2009). Student involvement and retention in higher education: The case for academic peer mentoring programmes for first-years. *Education as Change, 13*(1), 211–235. https://doi.org/10.1080/16823200909045077.
Loue, S. (2011). *Mentoring health science professionals*. New York: Springer.
Lumpe, A. T., & Staver, J. R. (1995). Peer collaboration and concept development: Learning about photosynthesis. *Journal of Research in Science Teaching: The Official Journal of the National Association for Research in Science Teaching, 32*(1), 71–98. https://doi.org/10.1002/tea.3660320108.
Mack, D. (2013). *Mentoring faculty of color: Essays on professional development and advancement in colleges and universities*. Jefferson, NC: McFarland & Company, Inc. Publishers.
MacLachlan, A. J. (2012). Minority undergraduate programs intended to increase participation in biomedical careers. *Mount Sinai Journal of Medicine: A Journal of Translational and Personalized Medicine, 79*(6), 769–781. https://doi.org/10.1002/msj.21350.
Malone, H. J. (2009). Build a bridge from high school to college: Transition programs are essential for many disadvantaged students. *Phi Kappa Phi Forum, 89*(3), 23.
Maltese, A. V., & Tai, R. H. (2011). Pipeline persistence: Examining the association of educational experiences with earned degrees in STEM among U.S. students. *Science Education, 95*(5), 877–907. https://doi.org/10.1002/sce.20441.
Manyibe, B. M., Manyibe, E. O., & Otiso, K. M. (2013). College student leadership development: An examination of pre-college leadership development of African students in the United States. *The Journal of Negro Education, 82*(4), 422–432. https://doi.org/10.7709/jnegroeducation.82.4.0422.
Maramba, D. C., & Museus, S. D. (2013). Examining the effects of campus climate, ethnic group cohesion, and cross-cultural interaction on Filipino American students' sense of belonging in college. *Journal of College Student Retention, 14*(4), 495–522.
Maramba, D. C., & Palmer, R. T. (2014). The impact of cultural validation on the college experiences of Southeast Asian American students. *Journal of College Student Development, 55*(6), 515–530. https://doi.org/10.1353/csd.2014.0054.
Martensen, B. F., Nykanen, D. K., Hart, M. C., & Bates, R. A. (2012). *Interdisciplinary STEM peer-mentoring and distance-based teams*. Paper presented at the American Society for Engineering Education.
Martin, R. H., & ebrary, I. (2008, 2010). *Racing Odysseus: A college president becomes a freshman again*. Berkeley, CA: University of California Press.

Martin, B. E., & Harris, F., III. (2006). Examining productive conceptions of masculinities: Lessons learned from academically driven African American male student-athletes. *The Journal of Men's Studies, 14*(3), 359–378. https://doi.org/10.3149/jms.1403.359.

Martin, F., Collier, K., & Carlon, S. (2009). Mentoring first-year distance education students in taxation studies. *Legal Education Review, 19*(1/2), 217–234.

Martin, J., Stefl, S., Cain, L., & Pfirman, A. (2020). Understanding first-generation undergraduate engineering students' entry and persistence through social capital theory. *International Journal of STEM Education, 7*(1). https://doi.org/10.1186/s40594-02000237-0.

Martindale, A. L., & Hammons, J. O. (2012). Some scholarship students need help, too: Implementation and assessment of a scholarship retention program. *Journal of College Student Retention, 14*(3), 371–395. https://doi.org/10.2190/CS.14.3.f.

Mavrinac, M. A. (2005). Transformational leadership: Peer mentoring as a values-based learning process. *Portal: Libraries and the Academy, 5*(3), 391–404. https://doi.org/10.1353/pla.2005.0037.

Maw, J. A. (2005). *The facilitation of student success: Incorporating affective, behavioural, and cognitive factors into first -year experience programs (Order No. NR08786). Available from ProQuest Dissertations & Theses Global. (305399879).*

Mayer, R., Christoffersen, R., & Fiorella, L. (2017). Enhancing undergraduate success in biology through the biomentors program. *American Biology Teacher, 79*(1), 21–25. https://doi.org/10.1525/abt.2017.79.1.23.

Mayhew, L. (1973). *The Carnegie Commission on Higher Education* (1st ed.). Jossey-Bass Publishers.

McAllister, G. F., & Irvine, J. J. (2000). Cross cultural competency and multicultural teacher education. *Review of Educational Research, 70*(1), 3–24.

McGee, R., Jr., Saran, S., & Krulwich, T. A. (2012). Diversity in the biomedical research workforce: Developing talent. *Mount Sinai Journal of Medicine: A Journal of Translational and Personalized Medicine, 79*(3), 397–411. https://doi.org/10.1002/msj.21310.

McGee, E., & Spencer, M. B. (2015). Black parents as advocates, motivators, and teachers of mathematics. *The Journal of Negro Education, 84*(3), 473–490. https://doi.org/10.7709/jnegroeducation.84.3.0473.

McGuire, K. M., Berhanu, J., Davis, C. H. F., & Harper, S. R. (2014). In search of progressive black masculinities: Critical self-reflections on gender identity development among black undergraduate men. *Men and Masculinities, 17*(3), 253–277. https://doi.org/10.1177/1097184X13514055.

McIntosh, A., Gidman, J., & Smith, D. (2014). Mentors' perceptions and experiences of supporting student nurses in practice. *International Journal of Nursing Practice, 20*(4), 360–365. https://doi.org/10.1111/ijn.12163.

McLoughlin, L. A. (2009). Success, recruitment, and retention of academically elite women students without STEM backgrounds in US undergraduate engineering education. *Engineering Studies, 1*(2), 151–168. https://doi.org/10.1080/19378620902911592.

McPhail, C. J., Robinson, M., & Scott, H. (2008). The cohort leadership development model: Student perspectives. *Community College Journal of Research and Practice, 32*(4-6), 362–374. https://doi.org/10.1080/10668920701884539.

McSherry, J. (2005). Challenges persist for minorities and women. *Electronic Design, 53*(23), 59–61.

Meacham, S. J. (2000). Black self-love, language, and the teacher education dilemma: The cultural denial and cultural limbo of African American preservice teachers. *Urban Education, 34*(5), 571–596. https://doi.org/10.1177/0042085900345003.

Medsker, L. L., & Trent, J. W. (1965). *The influence of different types of public higher institutions on college attendance from varying socioeconomic and ability levels*. Berkeley, CA: Center for the Study of Higher Education, University of California.

Merisotis, J. P., & Kee, A. M. (2006). A model of success: The model institutions for excellence program's decade of leadership in STEM education. *Journal of Hispanic Higher Education, 5*(3), 288–308. https://doi.org/10.1177/1538192706288422.

Mervis, J. (2009a). STEM education web site matches US scientists with teachers looking for help. *Science, 326*(5958), 1332.

Mervis, J. (2009b). "US STEM education" Obama's science advisers look at reform of schools. *Science, 326*(5953), 654.

Mervis, J. (2009c). US STEM education report calls for grassroots but comprehensive changes. *Science, 324*(5934), 1498.

Mervis, J. (2009d). NSF Boosts Success Rates, But at What Price? *Science (American Association for the Advancement of Science), 326* (5957), 1181–1182. https://doi.org/10.1126/science.326.5957.1181.

Mervis, J. (2013). US graduate education minorities run up significant debt in earning STEM Ph.Ds. *Science (American Association for the Advancement of Science), 340*(6140), 1510–1511. https://doi.org/10.1126/science.340.6140.1510.

Micari, M., Gould, A. K., & Lainez, L. (2010). Becoming a leader along the way: Embedding leadership training into a large-scale peer-learning program in the STEM disciplines. *Journal of College Student Development, 51*(2), 218–230. https://doi.org/10.1353/csd.0.0125.

Michael, J. (2006). Where's the evidence that active learning works? *American Journal of Physiology. Advances in Physiology Education, 30*(4), 159–167. https://doi.org/10.1152/advan.00053.2006.

Michael, J. A., & Modell, H. I. (2003). *Active learning in secondary and college science classrooms: A working model for helping the learner to learn.* Mahwah, NJ: L. Erlbaum Associates.

Mierson, S. (1998). A problem-based learning course in physiology for undergraduate and graduate basic science students. *Advances in Physiology Education, 275*(6), S16. https://doi.org/10.1152/advances.1998.275.6.S16.

Mihelich, J. A., Sarathchandra, D., Hormel, L., Storrs, D. A., & Wiest, M. M. (2016, 2017). Public understanding of science and K-12 STEM education outcomes: Effects of Idaho parents' orientation toward science on students' attitudes toward science. *Bulletin of Science, Technology & Society, 36*(3), 164-178. https://doi.org/10.1177/0270467616687217.

Miller-Cribbs, J. E., Cronen, S., Davis, L., & Johnson, S. D. (2002). An exploratory analysis of factors that foster school engagement and completion among African American students. *Children & Schools, 24*(3), 159.

Mills, M. T. (2010). Tools of engagement: Success course influence on student engagement. *Journal of the First-Year Experience & Students in Transition, 22*(2), 9.

Mills, C., Heyworth, J., Rosenwax, L., Carr, S., & Rosenberg, M. (2009). Factors associated with the academic success of first year health science students. *Advances in Health Sciences Education, 14*(2), 205–217. https://doi.org/10.1007/s10459-008-9103-9.

Mingle, J. R. (1981). *Challenges of retrenchment* (1st ed.). San Francisco, CA: Jossey-Bass.

Minority student leadership program begins peer mentoring: 29 students selected for participation. (2002). *ASHA Leader, 7*(2), 9.

Miranda, V. (2011). *Effectiveness of peer mentoring and college success courses on developing the self-efficacy of first-year community college students. ProQuest Dissertations Publishing.*

Mitchell, B. O. (2013). *Relationship of peer mentoring to academic success and social engagement for first year college students.* (Order No. 3609958). Available from ProQuest Dissertations & Theses Global. (1497943958).

Mizell, S., & Brown, S. (2016). The current status of STEM education research 2013-2015. *Journal of STEM Education: Innovations and Research, 17*(4), 52.

Moakler, M. W., & Kim, M. M. (2014). College major choice in STEM: Revisiting confidence and demographic factors. *The Career Development Quarterly, 62*(2), 128–142. https://doi.org/10.1002/j.2161-0045.2014.00075.x.

Mohr, N. (1998). Creating effective study groups for principals. *Educational Leadership, 55*(7), 41–44.

Mohr, J. J., Eiche, K., & Sedlacek, W. E. (1998). So close, yet so far: Predictors of attrition in college seniors. *Journal of College Student Development, 39*(4), 343–354.

Monaco, M., & Martin, M. (2007). The millennial student: a new generation of learners. *Athletic Training Education Journal, 2*(2), 42–46. https://doi.org/10.4085/1947-380X-2.2.42.

Mondisa, J. (2014). Designing sustainable mentoring programs: *Examining the role of social community in the STEM college student experience*. (Order No. 1585373). Available from ProQuest Dissertations & Theses Global. (1665302003).

Monitoring the graduation rate gap. (2008). *Diverse Issues in Higher Education, 25*(11), 16.

Montgomery, G. T., & Hirth, M. A. (2011). Freshman transition for at-risk students: Living with HEART. *NASSP Bulletin, 95*(4), 245–265. https://doi.org/10.1177/0192636511426618.

Morales, E. E., Ambrose-Roman, S., & Perez-Maldonado, R. (2016). Transmitting success: Comprehensive peer mentoring for at-risk students in developmental math. *Innovative Higher Education, 41*(2), 121–135. https://doi.org/10.1007/s10755-015-9335-6.

More than 100 higher education institutions utilize predictive analytics to raise graduation rates: Schools improving student outcomes through education advisory board's student success collaborative. (2014, Jun 10). PR Newswire.

Morgan, M. (2014). Study expectations of 1st/2nd generation STEM postgraduate taught students. *Quality Assurance in Education, 22*(2), 169–184. https://doi.org/10.1108/QAE-03-2013-0014.

Morris, D. S. (2015). Actively closing the gap? social class, organized activities, and academic achievement in high school. *Youth & Society, 47*(2), 267–290. https://doi.org/10.1177/0044118X12461159.

Morris, V. G., & Morris, C. L. (2002). Caring—The missing cin teacher education: Lessons learned from a segregated African American school. *Journal of Teacher Education, 53*(2), 120–123. https://doi.org/10.1177/0022487102053002005.

Morrow, R. M. (1994a). Issues facing engineering education. *Journal of Engineering Education, 83*(1), 15–18. https://doi.org/10.1002/j.2168-9830.1994.tb00109.x.

Morrow, W. (1994b). Entitlement and achievement in education. *Studies in Philosophy and Education, 13*(1), 33–47. https://doi.org/10.1007/BF01074084.

Moss, H. (2014). *A freshman survival guide for college students with autism spectrum disorders: The stuff nobody tells you about*. Philadelphia, PA: Jessica Kingsley Publishers.

Mullen, C. A. (2010). Editor's overview: Peer mentoring and collaborative discourses in higher education. *Mentoring and Tutoring: Partnership in Learning, 18*(2), 87–90. https://doi.org/10.1080/13611261003678846.

Murphy, M. J., Seneviratne, R. D., Cochrane, L., Davis, M. H., & Mires, G. J. (2013). Impact of student choice on academic performance: Cross-sectional and longitudinal observations of a student cohort. *BMC Medical Education, 13*(1), 26–26. https://doi.org/10.1186/1472-6920-13-26.

Museus, S. D. (2011). Generating ethnic minority student success (GEMS): A qualitative analysis of high-performing institutions. *Journal of Diversity in Higher Education, 4*(3), 147–162. https://doi.org/10.1037/a0022355.

National Center for Education Statistics. (2010a). *The nation's report card: Grade 12 Reading and Mathematics 2009 National and Pilot State results* (NCES 2011–455). National Center for Education Statistics, Institute of Education Sciences, U.S. Department of Education, Washington, D.C.

National Center for Education Statistics. (2010b). Integrated postsecondary education data system (IPEDS) compendium tables: 2010 ASI 4842-2.

National Research Council (U.S.), Board on Science Education, Center for Education, Committee on the Evaluation Framework for Successful K-12 STEM Education, & Board on Testing and Assessment. (2013, 2012). *Monitoring progress toward successful K-12 STEM education: A nation advancing?* Washington, DC: National Academies Press.

National Science Board (NSB). (2010). Science and Engineering Indicators (2010). Pages 2–29–2-30. NSB 10–01. Arlington, VA: National Science Foundation. Available at http://www.nsf.gov/statistics/seind/.

Naylor, R., Chakravarti, S., & Baik, C. (2016). Differing motivations and requirements in PhD student cohorts: A case study. *Issues in Educational Research, 26*(2), 351–367.

Nealy, M. (2009). Report: College graduation rates could dramatically decline. *Diverse Issues in Higher Education, 26*(11), 8.

Nelson, P. D. (2007). Striving for competence in the assessment of competence: Psychology's professional education and credentialing journey of public accountability. *Training and Education in Professional Psychology, 1*(1), 3–12. https://doi.org/10.1037/1931-3918.1.1.3.

Nelson, E. (2009). Copyright and distance education: The impact of the technology, education, and copyright harmonization act. *AACE Journal, 17*(2), 83–101.

Nelson, C. D. (2015). *Relationship between first-generation college students' expectations for experiences with faculty members and students' success after the first year*. ProQuest Dissertations Publishing.

Nerio, R., Webber, A., MacLachlan, E., Lopatto, D., & Caplan, A. (2019). One-Year Research Experience for Associate's Degree Students Impacts Graduation, STEM Retention, and Transfer Patterns. *CBE Life Sciences Education, 18*(2), ar25. https://doi.org/10.1187/cbe.19-02-0042.

New report shows graduation rate progress for minority students, but blacks trail other groups. (2015). *Journal of Blacks in Higher Education*. Retrieved from http://asu.summon.serialssolutions.com/2.0.0/link/0/eLvHCXMwzV1LS8QwEB58XATxgYqPFQa8WndrErs9iKi47mVRX-M8u2SalBWl108U_549zkjb4WN2z50LKkOSbL5Mv8wGw05NO8AMTtEhUKqWiB-HXG5JhJIWlxaZ5SNqa9qr_fzIB3S2tm24OkQ25VJrZo3iYqzSNhXVkvXl4DayNlr1u9p8b-fA3iFSPhVAdTWVVbkiXPxJgjMrHLTeTWoczq_iEVYJojvWsnggPdnYNtBbW8d3v2L-HifE_v1HM80e_0uMG7DWcGC8rBftJizoYgueCG6xPg7gMCvfDN5OpKobkeMDMWK8t1IyAmYkno2DvCit7x4O6-ac5hivphW6eqRBysD5M95Zcouuuma24ah383jdD3wAo2YX-mdFn-GwHVqWV_BeVexqodgHjSMeiM1ZahTEXkne5SKNYqpSFlGRlsgeteUPuz_98ACvEjYRTrkQtWKomU33ojIc-AG9R7xM

New U.S. distance education copyright act. (2002). *Copyright & New Media Law Newsletter, 6*(4), 8.

Newton, F. B., Ender, S. C. (2010). *Students helping students: A guide for peer educators on college campuses* (2nd ed.). San Francisco, CA: Jossey-Bass.

Nichols, L., & Islas, Á. (2016). Pushing and pulling emerging adults through college: College generational status and the influence of parents and others in the first year. *Journal of Adolescent Research, 31*(1), 59–95. https://doi.org/10.1177/0743558415586255.

Nickolich, D., Feldhaus, C., Cotton, S., Andrew, B., II, & Smallwood, J. (2010). Perceived life satisfaction of workplace specialist I faculty and mentors participating in a first-year STEM teacher training project. *Journal of Technology Studies, 36*(2), 41.

Noufou, O., Rezania, D., & Hossain, M. (2014). Measuring and exploring factors affecting students' willingness to engage in peer mentoring. *International Journal of Mentoring and Coaching in Education, 3*(2), 141–157. https://doi.org/10.1108/IJMCE-11-2012-0071.

Ntiri, D. W. (2001). Access to higher education for nontraditional students and minorities in a technology-focused society. *Urban Education, 36*(1), 129–144. https://doi.org/10.1177/0042085901361007.

O'Brien, M., Llamas, M., & Stevens, E. (2012). Lessons learned from four years of peer mentoring in a tiered group program within education. *Journal of the Australian and New Zealand Student Services Association, 40*, 7–15.

O'Connell, M. B., Jackson, A. N., Karaoui, L. R., Rodriguez de Bittner, M., Chen, A. M. H., Echeverri, M., & American College of Clinical Pharmacy. (2013). Cultural competency in health care and its implications for pharmacy part 3B: Emphasis on pharmacy education policy, procedures, and climate. *Pharmacotherapy: The Journal of Human Pharmacology and Drug Therapy, 33*(12), e368–e381. https://doi.org/10.1002/phar.1352.

Olivero, O. (2014). *Interdisciplinary mentoring in science: Strategies for success*. Elsevier.

Olson, S. (2012). *Engage to excel: Producing one million additional college graduates with degrees in science, technology, engineering, and mathematics*. Washington, DC: PCAST.

Olson, S., Labov, J. B., National Research Council (U.S.), Board on Science Education, Board on Life Sciences, Division of Behavioral and Social Sciences and Education (2012a). *Community colleges in the evolving STEM education landscape: Summary of a summit*. Washington, DC: National Academies Press.

Olson, S., Labov, J. B., National Research Council (U.S.), & Planning Committee on Evolving Relationships and Dynamics Between Two- and Four-Year Colleges and Universities. (2012b). *Community colleges in the evolving STEM education landscape: Summary of a summit*. Washington, DC: National Academies Press.

Oman, R. (2008). Five years later: What has the TEACH act taught us? (technology, education and copyright harmonization act of 2002). *Landslides, 1*(1), 26.

Ong, M., Jaumot-Pascual, N., & Ko, L. (2020). Research literature on women of color in undergraduate engineering education: A systematic thematic synthesis. *Journal of Engineering Education (Washington, D.C.), 109*(3), 581–615. https://doi.org/10.1002/jee.20345.

Osa, J., Oliver, A., & Walker, T. (2013). Mentoring and other professional support for faculty in institutions of higher learning: A study report. In A. A. Howley & M. B. Trube (Eds.), *Mentoring for the professions: Orienting toward the future* (pp. 127–144). Charlotte, NC: Information Age Publishing, Inc. Osa, Jet.al.

Spaulding, D. T., Kennedy, J. A., Rózsavölgyi, A., & Colón, W. (2020). Outcomes for peer-based mentors in a university-wide STEM persistence program: A three-year analysis. *Journal of College Science Teaching, 49*(4), 30–36.

Owens, M. A. (2013). "The feeling's mutual": Student participation in leadership as a cooperative effort. *The Review of Higher Education, 36*(4), 435–462. https://doi.org/10.1353/rhe.2013.0053.

Packard, B. W. (2004). Mentoring and retention in college science: Reflections on the sophomore year. *Journal of College Student Retention, 6*(3), 289–300. https://doi.org/10.2190/RUKP-XGVY-8LG0-75VH.

Padilla, R. V. (Ed.). (2008). *Student success modeling: Elementary school to college*. LLC: Stylus Publishing.

Palmer, R. T., & Maramba, D. C. (2011). African American male achievement: Using a tenet of critical theory to explain the African American male achievement disparity. *Education and Urban Society, 43*(4), 431–450. https://doi.org/10.1177/0013124510380715.

Palmer, R. T., Wood, J. L. (2013). *Community colleges and STEM: Examining underrepresented racial and ethnic minorities*. New York: Routledge, Taylor & Francis Group.

Pancer, S. M., Hunsberger, B., Pratt, M. W., & Alisat, S. (2000). Cognitive complexity of expectations and adjustment to university in the first year. *Journal of Adolescent Research, 15*(1), 38–57. https://doi.org/10.1177/0743558400151003.

Pang, T. (2010). Strengthen math education, improve student retention. *The Chronicle of Higher Education, 56*(19).

Park, J., Kim, Y., Salazar, C., & Hayes, S. (2020). Student-faculty interaction and discrimination from faculty in STEM: The link with retention. *Research in Higher Education, 61*(3), 330–356. https://doi.org/10.1007/s11162-019-09564-w.

Parks, M., Solmon, M., & Lee, A. (2007). Understanding classroom teachers' perceptions of integrating physical activity: A collective efficacy perspective. *Journal of Research in Childhood Education, 21*(3), 316–328. https://doi.org/10.1080/02568540709594597.

Pattengale, J., & Schreiner, L. A. (2000a). Policies and practices to enhance sophomore success. In *Visible solutions for invisible students: Helping sophomores succeed* (pp. 31–47). Columbia, SC: National Resource Center for the First-Year Experience & Students in Transition.

Pattengale, J., & Schreiner, L. A. (2000b). *Visible solutions for invisible students: Helping sophomores succeed*. Columbia, SC: National Resource Center for the Freshmen Year Experience and Students in Transition.

Patterson, R., Crede, E., Hines, K., Aarons, T., Bile, J., Edmister, W., & Watford, B. (2010). *Evaluating a peer leadership model in a large scale peer mentoring program*. Paper presented at the American Society for Engineering Education.

Pedersen, S., & Liu, M. (2003). Teachers' beliefs about issues in the implementation of a student-centered learning environment. *Educational Technology Research and Development, 51*(2), 57. https://doi.org/10.1007/BF02504526.

Pelham, W. E., & Fabiano, G. A. (2008). Evidence-based psychosocial treatments for attention-deficit/hyperactivity disorder. *Journal of Clinical Child & Adolescent Psychology, 37*(1), 184–214. https://doi.org/10.1080/15374410701818681.

Peralta, C., Caspary, M., & Boothe, D. (2013). Success factors impacting Latina/o persistence in higher education leading to STEM opportunities. *Cultural Studies of Science Education, 8*(4), 905–918. https://doi.org/10.1007/s11422-013-9520-9.

Perna, L. W., Jones, A. (2013). *The state of college access and completion: Improving college success for students from underrepresented groups*. New York: Routledge.

Persisting racial gap in college student graduation rates. (2014). *Journal of Blacks in Higher Education (Online), Retrieved from http://login.ezproxy1.lib.asu.edu/login?url=https://www-proquest-com.ezproxy1.lib.asu.edu/scholarly-journals/persisting-racial-gap-college-student-graduation/docview/1625005487/se-2?accountid=4485*

Pervin, L. A., Reik, L. E., Dalrymple, W., & Princeton University. (1966). *The college dropout and the utilization of talent*. Princeton, NJ: Princeton University Press.

Phipps, R. A., Wellman, J. V., & Institute for Higher Education Policy. (2001). *Funding the infostructure: A guide to financing technology infrastructure in higher education*. Indianapolis, IN: Lumina Foundation for Education.

Pike, K. L. (1962). *With heart and mind: A personal synthesis of scholarship and devotion*. Grand Rapids, MI: Eerdmans.

Pike, G., & Robbins, K. (2019). Using panel data to identify the effects of institutional characteristics, cohort characteristics, and institutional actions on graduation rates. *Research in Higher Education, 61*(4), 485–509. https://doi.org/10.1007/s11162-019-09567-7.

Piña, A. A. (2009). Factors influencing the institutionalization of distance education in higher education. *The Quarterly Review of Distance Education, 9*(4), 427.

Pino, N. W., & Smith, W. L. (2004). African American students, the academic ethic, and GPA. *Journal of Black Studies, 35*(1), 113–131. https://doi.org/10.1177/0021934704264954.

Pinnell, M., Rowly, J., Preiss, S., Franco, S., Blust, R., & Beach, R. (2013). Bridging the gap between engineering design and PK-12 curriculum development through the use of the STEM education quality framework. *Journal of STEM Education: Innovations and Research, 14*(4), 28.

Piper, J. K., & Krehbiel, D. (2015). Increasing STEM enrollment using targeted scholarships and an interdisciplinary seminar for first- and second-year college students. *Journal of STEM Education: Innovations and Research, 16*(4), 36.

Pittinsky, T. L., & Diamante, N. (2015). Going beyond fun in STEM: Instead of telling students that STEM is fun, tell them it's hard, and they'll feel a sense of pride and accomplishment at their success. *Phi Delta Kappan, 97*(2), 47.

Plotkin, H. (2002). All hail creative commons/Stanford professor and author Lawrence Lessig plans a legal insurrection. *San Francisco Chronicle.*

Poisel, M., & Joseph, S. (2011). Transfer students in higher education: Building foundations for policies, programs, and services that foster student success. *The first-year experience monograph series no. 54*. National Resource Center for The First-Year Experience and Students in Transition, University of South Carolina.

Poling, K. (2015). MySci advisors: Establishing a peer-mentoring program for first year science student support. *Collected Essays on Learning and Teaching, 8*, 181.

Pololi, L., & Evans, A. (2015). Group peer mentoring: An answer to the faculty mentoring problem? A successful program at a large academic department of medicine. *Journal of Continuing Education in the Health Professions, 35*(3), 192–200. https://doi.org/10.1002/chp.21296.

Power, R. K., Miles, B. B., Peruzzi, A., & Voerman, A. (2011). Building bridges: A practical guide to developing and implementing a subject-specific peer-to-peer academic mentoring program for first-year higher education students. *Asian Social Science, 7*(11), 75–80. https://doi.org/10.5539/ass.v7n11p75.

President's Council of Advisors on Science Technology, and President's Committee of Advisors on Science Technology. Report to the President, Engage to Excel Producing One Million Additional College Graduates with Degrees in Science, Technology, Engineering, and Mathematics. Washington, D.C.: Executive Office of the President, President's Council of Advisors on Science and Technology, 2012. Web.

Prendergast, L. Q. (2013). *Retention, success, and satisfaction of engineering students based on the first-year experience* (Order No. 3558652). Available from ProQuest Dissertations & Theses Global. (1344019689).

Price, D. B., Hyle, A. E., & Jordan, K. V. (2009). Ties that blind: Perpetuation of racial comfort and discomfort at a community college. *Community College Review, 37*(1), 3–33. https://doi.org/10.1177/0091552109336357.

Pritchard, P. A., & Grant, C. (2015a). *Success strategies for women in STEM: A portable mentor* (2nd ed.). Amsterdam: Academic Press.

Pritchard, P. A., & Grant, C. (2015b). *Success strategies from women in STEM: A portable mentor* (2nd ed.). Boston, MA: Academic Press.

Pritchard, P. A., & Grant, C. (2015c, 2014). *Success strategies for women in STEM: A portable mentor* (2nd ed.). Amsterdam: Academic Press.

Procter, C. (2012). Peer mentoring to secure student placements. *Higher Education, Skills and Work-Based Learning, 2*(2), 121–131. https://doi.org/10.1108/20423891211224603.

Pryce, J., Giovannetti, S., Spencer, R., Elledge, L. C., Gowdy, G., Whitley, M. L., & Cavell, T. A. (2015). Mentoring in the social context: Mentors' experiences with mentees' peers in a site-based program. *Children and Youth Services Review, 56*, 185–192. https://doi.org/10.1016/j.childyouth.2015.06.015.

Psychological factors in community college student retention. (2015). *Community College Journal of Research and Practice, 39*(3), 222-234. https://doi.org/10.1080/10668926.2013.803940.

Purves, A. (2001). *Pastoral theology in the classical tradition* (1st ed.). Louisville, KY; London: Westminster John Knox Press.

Quadlin, N. Y. (2016). Gender and time use in college: Converging or diverging pathways? *Gender & Society, 30*(2), 361–385. https://doi.org/10.1177/0891243215599648.

Quezada Sanders, I. G. (2013). *In the shadows: The invisible student cohort of Mexican diaspora: A phenomenological study of los retornos in Michoacán, México.*

Quintero, A. M. (2015). *Purposeful pathways to first-year student success: Examining integrative high impact practices at a Hispanic-serving institution.* (Order No. 3688316). Available from ProQuest Dissertations & Theses Global. (1673895474).

Racial gap in college student graduation rates. (2012). *Journal of Blacks in Higher Education (Online), Retrieved from http://login.ezproxy1.lib.asu.edu/login?url=https://www-proquest-com.ezproxy1.lib.asu.edu/scholarly-journals/racial-gap-college-student-graduation-rates/docview/1113333015/se-2?accountid=4485*

The racial gap in college graduation rates. (2014). *Journal of Blacks in Higher Education*. Retrieved from http://asu.summon.serialssolutions.com/2.0.0/link/0/eLvHCXMwzV1LSwMxEB58XATxgYqPCgFvwtrGJE1zKCJi24sX8V6mO1nay1rr-u_645ykG6pWe_ackAeZfDOZfMkHoG5vWtkPTPAmpwKR2EG1FY4UGmTj8rpg-b8x71X-_mYGkllavdgLJiNz0moekeVNqazlYN213N33LgoxUuG5Nmhp_N5AYIvIrA6jpq3E5yaOKN0PgODA3o1YDdfn8YjZhmyG-EyiDT3qwAtsRcnv7ME8veiIR-_eOVj57_C9zPIC9OgYW9wujPYQNXx7BNRuseMaQwRd9nIpJKeqkhujPkBY_knMFDo2P4ar3-PIwyNJghvWOeB8up6JOYBcDfb-s4jM_OgXhrHemNSJP0mmDuqNNYR1SoSQ7TMzPoLGuyfP1xReww3FOYNtk0jZgq5p9-MsoIvQJ0lvbvA

Radcliffe, R., & Bos, B. (2011). Mentoring approaches to create a college-going culture for at-risk secondary level students. *American Secondary Education, 39*(3), 86–107.

Radford, S. J. (2014). Building a mentor programme: Social networking can enhance communication among peer mentors in nursing education, writes Shellie Jean Radford. *Nursing Standard, 28*(39), 64.

Ramsay-Jordan, N. (2020). Hidden figures: How pecuniary influences help shape STEM experiences for Black students in Grades K-12. *Journal of Economics, Race, and Policy, 3*(3), 180–194. https://doi.org/10.1007/s41996-019-00049-7.

Rask, K. (2010). Attrition in STEM fields at a liberal arts college: The importance of grades and pre-collegiate preferences. *Economics of Education Review, 29*(6), 892–900. https://doi.org/10.1016/j.econedurev.2010.06.013.

Rawson, R. E., & Quinlan, K. M. (2002). Evaluation of a computer-based approach to teaching acid/base physiology. *Advances in Physiology Education, 26*(2), 85–97. https://doi.org/10.1152/advan.00042.2001.

Reauthorizing the Higher Education Act: access and innovation: hearing of the Committee on Health, Education, Labor, and Pensions, United States Senate, One Hundred Fifteenth Congress, second session, on examining reauthorizing the Higher Education Act, focusing on access and innovation, January 25, 2018. (2020). U.S. Government Publishing Office.

Redfield, R., Linton, R., & Herskovits, M. J. (1936). Memorandum for the study of acculturation. *American Anthropologist, 38*(1), 149–152. https://doi.org/10.1525/aa.1936.38.1.02a00330.

Rendón, L. I., Jalomo, R. E., & Nora, A. (2000). Theoretical considerations in the study of minority student retention in higher education. *Reworking The Student Departure Puzzle, 1*, 127–156.

Renn, K. A., & Arnold, K. D. (2003). Reconceptualizing research on college student peer culture. *The Journal of Higher Education, 74*(3), 261–291. https://doi.org/10.1080/00221546.2003.11780847.

Reyes, M. (2011). Unique challenges for women of color in STEM transferring from community colleges to universities. *Harvard Educational Review, 81*(2), 241–262.

Reynolds, L., Fisher, D., & Cavil, J. K. (2012). Impact of demographic variables on African-American student athletes' academic performance. *Educational Foundations, 26*(3-4), 93.

Rezai-Rashti, G., & Solomon, P. (2008). Teacher candidates' racial identity formation and the possibilities of antiracism in teacher education. In J. Zajda, L. Davies, & S. Majhanovich (Eds.), *Comparative and global pedagogies. Globalisation, comparative education and policy research* (pp. 167–187). Dordrecht, the Netherlands: Springer. https://doi.org/10.1007/978-1-4020-8349-5_10.

Rickards, W. H., Abromeit, J., Mentkowski, M., & Mernitz, H. (2016). Engaging faculty in an evaluative conversation. *New Directions for Evaluation, 2016*(151), 53–68. https://doi.org/10.1002/ev.20198.

Rickes, P. C. (2009). Make way for millennials! How today's students are shaping higher education space. *Planning for Higher Education, 37*(2), 7. Retrieved from https://www.scup.org/page/phe/read/article?data_id=29260&view=article.

Rickford, J. (1997). Suite for ebony and phonics. *Discover (Chicago, IL), 18*(12), 82.

Riegle-Crumb, C., Peng, M., & Russo-Tait, T. (2020). Committed to STEM? Examining factors that predict occupational commitment among Asian and White female students completing STEM US postsecondary programs. *Sex Roles, 82*(1-2), 102–116. https://doi.org/10.1007/s11199-019-01038-8.

Rieske, L. J., & Benjamin, M. (2015). Utilizing peer mentor roles in learning communities. *New Directions for Student Services, 2015*(149), 67–77. https://doi.org/10.1002/ss.20118.

Rios-Ellis, B., Rascón, M., Galvez, G., Inzunza-Franco, G., Bellamy, L., & Torres, A. (2015). Creating a model of Latino peer education: Weaving cultural capital into the fabric of academic services in an urban university setting. *Education and Urban Society, 47*(1), 33–55. https://doi.org/10.1177/0013124512468006.

Rising black student graduation rates at flagship state universities. (2014). *Journal of Blacks in Higher Education (Online), Retrieved from http://login.ezproxy1.lib.asu.edu/login?url=https://www-proquest-com.ezproxy1.lib.asu.edu/scholarly-journals/rising-black-student-graduation-rates-at-flagship/docview/1628966970/se-2?accountid=4485*

Roberts, D., & Ullom, C. (1989). Student leadership program model. *NASPA Journal, 27*(1), 67.

Robinson, L., Harris, A., & Burton, R. (2015). Saving face: Managing rapport in a problem-based learning group. *Active Learning in Higher Education, 16*(1), 11–24. https://doi.org/10.1177/1469787415573355.

Roscoe, J. L. (2015). Advising African American and Latino students. *Research and Teaching in Developmental Education, 31*(2), 48.

Ross, D., Hoppey, D., Halsall, S., McCallum, C., Hayes, S., & Hudson, R. (2005). Cohort use in teacher education: Benefits, barriers, and proposed solutions. *Teacher Education and Practice, 18*(3), 265.

Rotgans, J. I., & Schmidt, H. G. (2011). Situational interest and academic achievement in the active-learning classroom. *Learning and Instruction, 21*(1), 58–67. https://doi.org/10.1016/j.learninstruc.2009.11.001.

Rothstein, R. (2004a). *The achievement gap: A broader picture*. Alexandria, Egypt: Association for Supervision and Curriculum Development.

Rothstein, R. (2004b). The achievement gap: A broader picture. *Educational Leadership, 62*(3), 40–43.

Rothstein, R. (2004c). *Class and schools: Using social, economic, and educational reform to close the black-white achievement gap*. New York: Teachers College, Columbia University.

Rothstein, R. (2015). The racial achievement gap, segregated schools, and segregated neighborhoods: A constitutional insult. *Race and Social Problems, 7*(1), 21–30. https://doi.org/10.1007/s12552-014-9134-1.

Rubin, M. (2008). Gap year: Transition from high school to college. Time off or time on? *Hispania, 91*(1), 225–226.

Ruland, J., & Gafni Lachter, L. (2015). Enhancing leadership and communication by utilizing a peer-mentoring program: Lessons from three years of implementation. *American Journal of Occupational Therapy, 69*(Suppl. 1), 6911510134. https://doi.org/10.5014/ajot.2015.69S1-PO4079.

Russell, C., & ERIC Clearinghouse on Information and Technology. (2002). *New copyright exemptions for distance educators: The technology, education and copyright harmonization (TEACH) act*. Syracuse, NY: ERIC Clearinghouse on Information and Technology.

Sadler, P. M., Sonnert, G., Hazari, Z., & Tai, R. (2014). The role of advanced high school coursework in increasing STEM career interest. *Science Educator, 23*(1), 1.

Sáenz, V. B., Ponjuan, L., Segovia, J., & Del Real Viramontes, J. (2015). Developing a Latino mentoring program: Project MALES (mentoring to achieve Latino educational success). *New Directions for Higher Education, 2015*(171), 75–85. https://doi.org/10.1002/he.20144.

Salazar, M. J., Pazey, B. L., & Zembik, M. K. (2013). What we've learned and how we've used it: Learning experiences from the cohort of a high-quality principalship program. *Journal of Research on Leadership Education, 8*(3), 304.

Saltmarsh, J. A., Zlotkowski, E. A., & ebrary, I. (2011). *Higher education and democracy: Essays on service-learning and civic engagement*. Philadelphia, PA: Temple University Press.

Salto, L. M., Riggs, M. L., Leon, D. D. D., Casiano, C. A., & Leon, M. D. (2014). Underrepresented minority high school and college students report STEM-pipeline sustaining gains after participating in the Loma Linda University summer health disparities research program: E108497. *PLoS One, 9*(9). https://doi.org/10.1371/journal.pone.0108497.

Same-ethnicity mentors boost students' commitment to STEM careers. (2012, May 6). NewsRx Health & Science, 231. https://link.gale.com/apps/doc/A288912247/CWI?u=asuniv&sid=CWI&xid=8f2799a8

Sanchez, R. J., Bauer, T. N., & Paronto, M. E. (2006). Peer-mentoring freshmen: Implications for satisfaction, commitment and retention to graduation. *Academy of Management Learning & Education, 5*(1), 25–37. https://doi.org/10.5465/AMLE.2006.20388382.

Sanchez-Leguelinel, C. (2008). Supporting 'slumping' sophomores: Programmatic peer initiatives designed to enhance retention in the crucial second year of college. *College Student Journal, 42*(2), 637.

Saunders-Smits, G., Nieweg, M., & ebrary, I. (2005). *Research and practice of active learning in engineering education*. Amsterdam: Pallas Pub./Amsterdam University Press.

Savery, J. R., & Duffy, T. M. (1995). Problem based learning: An instructional model and its constructivist framework. *Educational Technology, 35*(5), 31–38. Retrieved from https://www.jstor.org/stable/44428296?seq=1#page_scan_tab_contents.

Sax, L. J., Allison Kanny, M., Jacobs, J. A., Whang, H., Weintraub, D. S., & Hroch, A. (2016). Understanding the changing dynamics of the gender gap in undergraduate engineering majors: 1971–2011. *Research in Higher Education, 57*(5), 570–600. https://doi.org/10.1007/s11162-015-9396-5.

Schatz, G. (2004). "Letter to a young scientist" by Gottfried. *FEBS Letters, 558*. https://doi.org/10.1016/S0014-5793(03)01500-X.

Schmidt, E. K., & Faber, S. T. (2016). Benefits of peer mentoring to mentors, female mentees and higher education institutions. *Mentoring and Tutoring: Partnership in Learning, 24*(2), 137–157. https://doi.org/10.1080/13611267.2016.1170560.

Schmidt, H. G., Cohen-Schotanus, J., & Arends, L. R. (2009). Impact of problem-based, active learning on graduation rates for 10 generations of Dutch medical students. *Medical Education, 43*(3), 211–218. https://doi.org/10.1111/j.1365-2923.2008.03287.x.

Schneider, B., Broda, M., Judy, J., & Burkander, K. (2013). Pathways to college and STEM careers: Enhancing the high school experience. *New Directions for Youth Development, 2013*(140), 9–29. https://doi.org/10.1002/yd.20076.

Schneider, K. R., Bickel, A., & Morrison-Shetlar, A. (2015). Planning and implementing a comprehensive student-centered research program for first-year STEM undergraduates. *Journal of College Science Teaching, 44*(3), 37.

Schweingruber, H., Pearson, G., Honey, M., & National Academy of Engineering and National Research Council. (2014). *STEM integration in K-12 education: Status, prospects, and an agenda for research*. Washington, DC: National Academies Press.

Scott, P. (1975). *Strategies for postsecondary education*. New York: Wiley.

Scott, T. P., Milam, J. L., Stuessy, C. L., Blount, K. P., & Bentz, A. (2006). Math and science scholars (MASS) program: A model program for the recruitment and retention of preservice mathematics and science teachers. *Journal of Science Teacher Education, 17*(4), 389–411. https://doi.org/10.1007/s10972-006-9026-3.

Sega Enters. v. Accolade, Inc. (n.d.) No. 92-15655 (United States Court of Appeals for the Ninth Circuit).

Seidman, A. (2005a). *College student retention: Formula for student success*. Westport, CT: Praeger Publishers.

Seidman, A. (2005b). Minority student retention: Resources for practitioners. *New Directions for Institutional Research, 2005*(125), 7–24. https://doi.org/10.1002/ir.136.

Seidman, A., Astin, A., Berger, J., Bibo, E., Burkum, K., Cabrera, A., Crisp, G., Gansemer-Topf, A., Hagedorn, L., & LaNasa, S. (2012). College student retention: Formula for student success. In *College Student Retention*. Lanham MD: Rowman & Littlefield Publishers.

Sessions, W. L. (1994). *The concept of faith: A philosophical investigation*. Ithaca, NY: Cornell University Press.

Shaw, S., & Fairhurst, D. (2008). Engaging a new generation of graduates. *Education and Training, 50*(5), 366–378. https://doi.org/10.1108/00400910810889057.

Sheridan, R., Hammaker, D., de Peralta, T., & Fitzgerald, M. (2016). Dental students' perceived value of peer-mentoring clinical leadership experiences. *Journal of Dental Education, 80*(3), 311–317.

Shin, J. E. L., Levy, S. R., & London, B. (2016). Effects of role model exposure on STEM and non-STEM student engagement. *Journal of Applied Social Psychology, 46*(7), 410–427. https://doi.org/10.1111/jasp.12371.

Shuler, J. A. (2003). Distance education, copyrights rights, and the new TEACH act. *The Journal of Academic Librarianship, 29*(1), 49–51. https://doi.org/10.1016/S0099-1333(02)00386-5.

Sidle, M. W., & McReynolds, J. (2009). The freshman year experience: Student retention and student success. *NASPA Journal, 46*(3), 434. https://doi.org/10.2202/1949-6605.5019.

Silverman, L. H., & Seidman, A. (2011). Academic progress in developmental math courses: A comparative study of student retention. *Journal of College Student Retention, 13*(3), 267–287. https://doi.org/10.2190/CS.13.3.a.

Simonson, M. (2010). Retention. *Distance Learning (Greenwich, CT), 7*(3), 92.

Sin, J. H., Pathickal, S. M., & Li, M. (2015). Establishment of a peer-mentoring program for student pharmacists. *American Journal of Health-System Pharmacy, 72*(19), 1610–1611. https://doi.org/10.2146/ajhp140544.

Singh, S., Singh, N., & Dhaliwal, U. (2014). Near-peer mentoring to complement faculty mentoring of first-year medical students in India. *Journal of Educational Evaluation for Health Professions, 11*, 12. https://doi.org/10.3352/jeehp.2014.11.12.

Singleton-Jackson, J. A., Jackson, D. L., & Reinhardt, J. (2010). Students as consumers of knowledge: Are they buying what we're selling? *Innovative Higher Education, 35*(5), 343. https://doi.org/10.1007/s10755-010-9151-y.

Skaniakos, T., Penttinen, L., & Lairio, M. (2014). Peer group mentoring programmes in Finnish higher education – mentors' perspectives. *Mentoring & Tutoring: Partnership in Learning, 22*(1), 74–86. https://doi.org/10.1080/13611267.2014.882609.

Skelton, C., & Francis, B. (2011). Successful boys and literacy: Are "literate boys" challenging or repackaging hegemonic masculinity? *Curriculum Inquiry, 41*(4), 456–479. https://doi.org/10.1111/j.1467-873X.2011.00559.x.

Skoglund, F., Ness, J. (2011). *Student success: How to make it happen*. ProQuest Ebook Central.

Slovacek, S. P., Whittinghill, J. C., Tucker, S., Rath, K. A., Peterfreund, A. R., Kuehn, G. D., & Reinke, Y. G. (2011). Minority students severely underrepresented in science, technology engineering and math. *Journal of STEM Education: Innovations and Research, 12*(1), 5–16.

Smith, B. (2013). *Mentoring at-risk students through the hidden curriculum of higher education*. Lanham, MD: Lexington Books.

Smith, C. H., & Jiménez, M. (2014). Razing the bar: Developmental students shattering expectations in a first-year learning community. *Teaching English in the Two Year College, 42*(1), 41.

Smith, T. R., McGowan, J., Allen, A. R., Johnson, W. D., Dickson, L. A., Najee-ullah, M. A., & Peters, M. (2008). Evaluating the impact of a faculty learning community on STEM teaching and learning. *The Journal of Negro Education, 77*(3), 203–226.

Smith, W. L., & Zhang, P. (2008). Perceived factors facilitating students' transition from high school to college. *Michigan Sociological Review*, 22, 19–40.

Snyder, J., Sloane, J., Dunk, R., & Wiles, J. (2016). Peer-led team learning helps minority students succeed. *PLoS Biology, 14*(3), e1002398. https://doi.org/10.1371/journal.pbio.1002398.

Sony Corp. of Am. v. Universal City Studios, Inc. (n.d.) No. 81-1687 (Supreme Court of the United States).

Soom, C. V., & Donche, V. (2014). Profiling first-year students in STEM programs based on autonomous motivation and academic self-concept and relationship with academic achievement. *PLoS One, 9*(11), E112489. https://doi.org/10.1371/journal.pone.0112489.

Sosik, J. (2000). Leadership styles, mentoring functions received, and job-related stress: A conceptual model and preliminary study. *Journal of Organizational Behavior, 21*(4), 365.

Soto, N., Cuadrado, A., Garcia, M., Risquez, A., & Suarez-Ortega, M. (2012). The role of mentor in higher education mentoring process. *Educacion xx1, 15*(2), 93–118.

Sottie, C. A., Dubus, N., & Sossou, M. (2013). Enhancing student outcomes through mentoring, peer counselling and parental involvement. *Prospects, 43*(3), 377–391. https://doi.org/10.1007/s11125-013-9280-z.

South Carolina Univ., Columbia. National Resource Center for the Freshman Year Experience and Students in Transition. (1996). A passport to enhanced student learning, success, satisfaction, and retention. *Proceedings of the International Conference on the First-Year Experience (9th, St. Andrews, Scotland, United Kingdom, July 15–19, 1996).*

Spector, N. D., Mann, K. J., Anderson, M. S., Narayan, A. P., & McGregor, R. (2010). Facilitated peer group mentoring: A case study of creating leadership skills among the associate program directors of the APPD. *Academic Pediatrics, 10*(3), 161–164. https://doi.org/10.1016/j.acap.2010.03.002.

Spence, J. T., Helmreich, R., & Stapp, J. (1975). Ratings of self and peers on sex role attributes and their relation to self-esteem and conceptions of masculinity and femininity. *Journal of Personality and Social Psychology, 32*(1), 29–39. https://doi.org/10.1037/h0076857.

Spencer, K. M. (2012). *A study of the impact of a first-year experience initiative on first-year developmental education student success and persistence*. (Order No. 3542047). Available from ProQuest Dissertations & Theses Global. (1145463048).

Spicer, Y. M. (2015). *Closing the achievement gap and building the pipeline through STEM education: A US perspective*. https://doi.org/10.4324/9781315714172

Spronken-Smith, R., & Harland, T. (2009). Learning to teach with problem-based learning. *Active Learning in Higher Education, 10*(2), 138–153. https://doi.org/10.1177/1469787409104787.

Stanley, C. A. (2006). Coloring the academic landscape: Faculty of color breaking the silence in predominantly white colleges and universities. *American Educational Research Journal, 43*(4), 701–736. https://doi.org/10.3102/00028312043004701.

Stanley, C. A. (2007). When counter narratives meet master narratives in the journal editorial-review process. *Educational Researcher, 36*(1), 14–24. https://doi.org/10.3102/0013189X06298008.

State of Black Arizona, Arizona Community Foundation, Arizona Public Service Company, & ASU Office of Public Affairs. (2012). The State of Black Arizona: Volume 3.

Steinert, Y., O'Sullivan, P., & Irby, D. (2019). Strengthening teachers' professional identities through faculty development. *Academic Medicine, 94*(7), 963–968. https://doi.org/10.1097/acm.0000000000002695.

Stephens, N. M., Hamedani, M. G., & Destin, M. (2014). Closing the social-class achievement gap: A difference-education intervention improves first-generation students' academic performance and all students' college transition. *Psychological Science, 25*(4), 943–953. https://doi.org/10.1177/0956797613518349.

Stewart, D. L., & Lozano, A. (2009). Difficult dialogues at the intersections of race, culture, and religion. *New Directions for Student Services, 2009*(125), 23–31. https://doi.org/10.1002/ss.304.

Stewart, M. F., & Chisholm, C. U. (2012). Comparative analysis of emotional competency within distinct student cohorts. *Global Journal of Engineering Education, 14*(2), 163–169.

Stewart, S. (2015). Factors influencing college persistence for first-time students. *Journal of Developmental Education, 38*(3), 12.

Stewart, T. A., Owens, D., Queener, J. E., & Reynolds, C. A. (2014). The relationship between racial identity and acculturative stress among African American students in counselor training programs. *Michigan Journal of Counseling, 41*(1), 4.

Stone, F. M. (2004). *The mentoring advantage: Creating the next generation of leaders*. Chicago: Dearborn Trade Publishing.

Strayhorn, T. L. (2015). Factors influencing black males' preparation for college and success in STEM majors: A mixed methods study. *The Western Journal of Black Studies, 39*(1), 45–45.

Strayhorn, T. L. (2012). *College students' sense of belonging: A key to educational success for all students*. New York; London: Routledge.

Stuart, R. (2009). Will they come? *Diverse Issues in Higher Education, 26*(16), 9–9.

Stuart, R. (2013). Dropout graduates: Under antiquated IPEDS measures, some graduates and transfers are still considered dropouts. *Diverse Issues in Higher Education, 30*(18), 13.

Stuber, J. M. (2011). *Inside the college gates: How class and culture matter in higher education*. Lanham, MD: Lexington Books.

Stupnisky, R. H., Stewart, T. L., Daniels, L. M., & Perry, R. P. (2011). When do students ask why? Examining the precursors and outcomes of causal search among first-year college students. *Contemporary Educational Psychology, 36*(3), 201–211. https://doi.org/10.1016/j.cedpsych.2010.06.004.

Su, X. (2013). International doctoral science and engineering students: Impact on cohorts' career prospects. *Journal of Studies in International Education, 17*(5), 590–606. https://doi.org/10.1177/1028315312467354.

Sullins, E. S., Hernandez, D., Fuller, C., & Tashiro, J. S. (1995). Predicting who will major in a science discipline: Expectancy-value theory as part of an ecological model for studying academic communities. *Journal of Research in Science Teaching, 32*(1), 99–119. https://doi.org/10.1002/tea.3660320109.

Svinicki, M. D. (1998). A theoretical foundation for discovery learning. *Advances in Physiology Education, 275*(6), S4. https://doi.org/10.1152/advances.1998.275.6.S4.

Syed, M., Goza, B. K., Chemers, M. M., & Zurbriggen, E. L. (2012). Individual differences in preferences for matched-ethnic mentors among high-achieving ethnically diverse adolescents in STEM. *Child Development, 83*(3), 896–910. https://doi.org/10.1111/j.1467-8624.2012.01744.x.

Szabo, A. (2015). Identity development during cultural transition: The role of social-cognitive identity processes. *International Journal of Intercultural Relations, 46*(13), 13–25.

Szelényi, K., Denson, N., & Inkelas, K. K. (2013). Women in STEM majors and professional outcome expectations: The role of living-learning programs and other college environments. *Research in Higher Education, 54*(8), 851–873. https://doi.org/10.1007/s11162-013-9299-2.

Talab, R. (2003). An initial look at the TEACH act. *TechTrends, 47*(2), 4–6. https://doi.org/10.1007/BF02763414.

Talmadge, L. (2011). *Mentoring programs help urban youth. In Opposing Viewpoints: Urban America. Retrieved from https://search.lib.asu.edu/permalink/01ASU_INST/fdcm53/cdi_gale_incontextnref_IOV_EJ3010399249*

Tandogan, R. O., & Orhan, A. (2007). The effects of problem-based active learning in science education on students' academic achievement, attitude and concept learning. *3*(1), 71.

Tawbush, R., Stanley, S., Campbell, T., & Webb, M. (2020). International comparison of K-12 STEM teaching practices. *Journal of Research in Innovative Teaching & Learning, 13*(1), 115–128. https://doi.org/10.1108/JRIT-01-2020-0004.

Tenenbaum, L. S., Anderson, M. K., Jett, M., & Yourick, D. L. (2014). An innovative near-peer mentoring model for undergraduate and secondary students: STEM focus. *Innovative Higher Education, 39*(5), 375–385. https://doi.org/10.1007/s10755-014-9286-3.

Terrion, J. (2012). Student peer mentors as a navigational resource in higher education In The SAGE Handbook of Mentoring and Coaching in Education (p. 383–). SAGE Publications Ltd. https://doi.org/10.4135/9781446247549.n26

Terrion, J. L. (2012). *Student peer mentors as a navigational resource in higher education*. London: SAGE Publications Ltd. https://doi.org/10.4135/9781446247549.n26.

Thakral, C., Vasquez, P., Bottoms, B., Matthews, A., Hudson, K., & Whitley, S. (2016). Understanding difference through dialogue: A first-year experience for college students. *Journal of Diversity in Higher Education, 9*(2), 130–142. https://doi.org/10.1037/a0039935.

Thiry, H., & Laursen, S. L. (2011). The role of student-advisor interactions in apprenticing undergraduate researchers into a scientific community of practice. *Journal of Science Education and Technology, 20*(6), 771–784. https://doi.org/10.1007/s10956-010-9271-2.

Thomas, J. C., Wolters, C., Horn, C., & Kennedy, H. (2014). Examining relevant influences on the persistence of African-American college students at a diverse urban university. *Journal of College Student Retention, 15*(4), 551–573.

Thomas, K. M., Hu, C., Gewin, A. G., Bingham, K., & Yanchus, N. (2005). The roles of protégé race, gender, and proactive socialization attempts on peer mentoring. *Advances in Developing Human Resources, 7*(4), 540–555. https://doi.org/10.1177/1523422305279681.

Thomas, N., Bystydzienski, J., & Desai, A. (2015). Changing institutional culture through peer mentoring of women STEM faculty. *Innovative Higher Education, 40*(2), 143–157. https://doi.org/10.1007/s10755-014-9300-9.

Tomlinson-Clarke, S. (1998). Dimensions of Adjustment among College Women. *Journal of College Student Development, 39*(4), 364–372.

Tierney, W. G., & Lechuga, V. M. (2004). *Restructuring shared governance in higher education*. San Francisco, CA: Jossey-Bass.

Tinto, V. (1993). *Leaving college: Rethinking the causes and cures of student attrition*. Chicago, IL: University of Chicago Press.

Tinto, V. (2012). *Leaving college: Rethinking the causes and cures of student attrition* (2nd ed.). University of Chicago Press.

Tinto, V. (1997). Classrooms as communities. *The Journal of Higher Education (Columbus), 68*(6), 599.

Tobin, J. J., Wu, D. Y. H., & Davidson, D. H. (1989). *Preschool in three cultures: Japan, China, and the United States*. New Haven, CT: Yale University Press.

Tobolowsky, B. F. (2008). Sophomores in transition: The forgotten year. *New Directions for Higher Education, 2008*(144), 59–67. https://doi.org/10.1002/he.326.

Tofel-Grehl, C., & Callahan, C. M. (2017). STEM high schools teachers' belief regarding STEM student giftedness. *The Gifted Child Quarterly, 61*(1), 40–51. https://doi.org/10.1177/0016986216673712.

Tomasko, D. L., Ridgway, J. S., Waller, R. J., & Olesik, S. V. (2016). Association of summer bridge program outcomes with STEM retention of targeted demographic groups. *Journal of College Science Teaching, 45*(4), 90.

Touchie, C., De Champlain, A., Pugh, D., Downing, S., & Bordage, G. (2014). Supervising incoming first-year residents: Faculty expectations versus residents' experiences. *Medical Education, 48*(9), 921–929. https://doi.org/10.1111/medu.12503.

Tremblay, P. F., & Rodger, S. (2003). The effects of a peer mentoring program on academic success among first year university students. *The Canadian Journal of Higher Education, 33*(3), 1.

Tsen, L. C., Borus, J. F., Nadelson, C. C., Seely, E. W., Haas, A., & Fuhlbrigge, A. L. (2012). The development, implementation, and assessment of an innovative faculty mentoring leadership program. *Academic Medicine, 87*(12), 1757–1761. https://doi.org/10.1097/ACM.0b013e3182712cff.

Tsouroufli, M., Rees, C. E., Monrouxe, L. V., & Sundaram, V. (2011). Gender, identities and intersectionality in medical education research. *Medical Education, 45*(3), 213–216. https://doi.org/10.1111/j.1365-2923.2010.03908.x.

Turner, P. (2014). College retention initiatives meeting the needs of millennial freshman students. *College Student Journal, 48*(1), 94.

Turner, P., & Thompson, E. (2014). College retention initiatives meeting the needs of millennial freshman students. *College Student Journal, 48*(1), 94.

Tytler, R., Freeman, B., Marginson, S. (2015). *The age of STEM: Educational policy and practice across the world in science, technology, engineering and mathematics*. New York; London: Routledge, Taylor & Francis Group.

UMG Recordings, Inc. v. MP3.Com, Inc. 00 Civ. 0472 (JSR) Lexis 17907 (United States District Court for the Southern District of New York 2000).

United States. National Commission on Excellence in Education. (1983). *A nation at risk: the imperative for educational reform*. Washington, DC: The National Commission on Excellence in Education.

United States. (2015). *STEM education act of 2015*. Washington, DC: U.S. Government Publishing Office.

United States, Senate, Committee on Commerce, Science, and Transportation, & Congress. (2015). *STEM education act of 2015: Report of the Committee on Commerce, Science and Transportation on H.R. 1020*. Washington: U.S. Government Publishing Office. Retrieved from http://asu.summon.serialssolutions.com/2.0.0/link/0/eLvHCXMwY2AwNtIz0EUrE8xNQAe1p1omm1iaGVkmJ6YmpxhZmqSaWJikgjZvmqPOzMBPzYENvCGt1YEPXpgDOyaGZubMDMzA4hXUDvIMQlwMAy3u3AQZWEB7BoQYmFLzhBkEYDclKEAzjgiDVHCIq68CfDGFgmNyiUJ-mgKwWjQVZZBxcw1x9tAFmhkPHUyJh1lraCTGwJsIWoSeVwLerJYiwaBgYZyaYpmYlGRkmWpikmRpammSnAasjo1MUkzTkk0SjSUZRLGaJYVDXJqBC-QQSI9fhoE1DZgiU2XBngQAr45e8Q

Unzueta, C. H., Moores-Abdool, W., & Vazquez Donet, D. (2010). Perceptions of special education professors and culturally linguistically diverse doctoral students on cohorts. *Teacher Education and Special Education, 33*(2), 169–182. https://doi.org/10.1177/0888406409360009.

Upcraft, M. L., & Gardner, J. N. (1989). *The freshman year experience: Helping students survive and succeed in college* (1st ed.). San Francisco, CA: Jossey-Bass Publishers.

Ural, A. (2013, 2010). A hands-on finite element modeling experience in a multidisciplinary project-based freshman course. *Computer Applications in Engineering Education, 21*(2), 294-299. https://doi.org/10.1002/cae.20471.

Urias, M. V. (2014). Black male graduation rates in community colleges: Do institutional characteristics make a difference. *Community College Journal of Research and Practice, 38*(12), 1112–1124. https://doi.org/10.1080/10668926.2012.745101.

USA funds awards complete college America $1.5 million grant to develop and deploy purpose first: New initiative will combine workforce connections, predictive analytics and career advising to empower students in choosing majors and careers. (2014). *NASDAQ OMX's News Release Distribution Channel.*

Uygun, N., & Tertemiz, N. (2014). Effects of problem-based learning on student attitudes, achievement and retention of learning in math course. *Egitim Ve Bilim-Education and Science, 39*(174), 75–90.

Valencia, R. R. (2015). *Students of color and the achievement gap: Systematic challenges, systematic transformations.* New York, NY: Routledge, Taylor & Francis Group.

Vantieghem, W., Vermeersch, H., & Van Houtte, M. (2014a). Why "gender" disappeared from the gender gap: (re-)Introducing gender identity theory to educational gender gap research. *Social Psychology of Education, 17*(3), 357–381. https://doi.org/10.1007/s11218-014-9248-8.

Vantieghem, W., Vermeersch, H., & Van Houtte, M. (2014b). Transcending the gender dichotomy in educational gender gap research: The association between gender identity and academic self-efficacy. *Contemporary Educational Psychology, 39*(4), 369–378. https://doi.org/10.1016/j.cedpsych.2014.10.001.

Van Wyk, J., & Rennie, C. (2015). Learning anatomy through dissection: Perceptions of a diverse medical student cohort. *International Journal of Morphology, 33*(1), 89–95.

Varma, R. (2002). Women in information technology: A case study of undergraduate students in a minority-serving institution. *Bulletin of Science, Technology & Society, 22*(4), 274–282. https://doi.org/10.1177/0270467602022004003.

Venezia, A., & Jaeger, L. (2013). Transitions from high school to college. *The Future of Children, 23*(1), 117–136.

Vesilind, P. A. (2001). Mentoring engineering students: Turning pebbles into diamonds. *Journal of Engineering Education, 90*(3), 407–411. https://doi.org/10.1002/j.2168-9830.2001.tb00620.x.

Villa, E. Q., Esquinca, A., Mein, E. L., Golding, P., & Hug, S. (2015). An engineering student leadership development model for 21st century engineering: Using sociocultural theory to inform practice. Proceedings of the 2015 IEEE Frontiers in Education Conference (FIE) IEEE Computer Society Washington, DC United States, ISBN 978-1-4799-8454-1. https://doi.org/10.1109/FIE.2015.7344034.

Villegas, A. M., & Lucas, T. (2002). *Educating culturally responsive teachers: A coherent approach.* Albany: State University of New York Press.

Van Waes, S., Van den Bossche, P., Moolenaar, N. M., De Maeyer, S., & Van Petegem, P. (2015). Know-who? Linking faculty's networks to stages of instructional development. *Higher Education, 70*(5), 807–826. https://doi.org/10.1007/s10734-015-9868-8.

Walker, D., & Verklan, T. (2016). Peer mentoring during practicum to reduce anxiety in first-semester nursing students. *Journal of Nursing Education, 55*(11), 651–654. https://doi.org/10.3928/01484834-20161011-08.

Walker, K., & Dixon, V. (2002). Spirituality and academic performance among African American college students. *Journal of Black Psychology, 28*(2), 107–120.

Wallace, S. (2014). Impact of a student leadership development program. *American Journal of Pharmaceutical Education, 78*(4), 1.

Walpole, M., McDonough, P. M., Bauer, C. J., Gibson, C., Kanyi, K., & Toliver, R. (2005). This test is unfair: Urban African American and Latino high school students' perceptions of standardized college admission tests. *Urban Education, 40*(3), 321–349. https://doi.org/10.1177/0042085905274536.

Walpole, M., Simmerman, H., Mack, C., Mills, J. T., Scales, M., & Albano, D. (2008). Bridge to success: Insight into summer bridge program students' college transition. *Journal of the First-Year Experience & Students in Transition, 20*(1), 11.

Walsh, B. D., Vacha-haase, T., Kapes, J. T., Dresden, J. H., Thomson, W. A., & Ochoa-Shargey, B. (1996). The values scale: Differences across grade levels for ethnic minority students. *Educational and Psychological Measurement, 56*(2), 263–275. https://doi.org/10.1177/0013164496056002007.

Wan, C., & Chew, P. Y. (2013). Cultural knowledge, category label, and social connections: Components of cultural identity in the global, multicultural context. *Asian Journal of Social Psychology, 16*(4), 247–259. https://doi.org/10.1111/ajsp.12029.

Wang, V. (2017). *Theory and practice of adult and higher education*. Charlotte, NC: Information Age Publishing, Incorporated.

Wang, X. (2013a). Modeling entrance into STEM fields of study among students beginning at community colleges and four-year institutions. *Research in Higher Education, 54*(6), 664–692. https://doi.org/10.1007/s11162-013-9291-x.

Wang, X. (2013b). Why students choose STEM majors: Motivation, high school learning, and postsecondary context of support. *American Educational Research Journal, 50*(5), 1081–1121. https://doi.org/10.3102/0002831213488622.

Wang, X. (2015, 2014). Pathway to a baccalaureate in STEM fields: Are community colleges a viable route and does early STEM momentum matter? *Educational Evaluation and Policy Analysis, 37*(3), 376-393. https://doi.org/10.3102/0162373714552561.

Wang, Y. (2020). Design-based research on integrating learning technology tools into higher education classes to achieve active learning. *Computers in Education, 156*, 103935. https://doi.org/10.1016/j.compedu.2020.103935.

Ward, C., Jones, K. W., Coles, R., Rich, L., Knapp, S., & Madsen, R. (2014a). Mentored research in a tribal college setting: The Northern Cheyenne case. *Journal of Research in Rural Education, 29*(3), 1.

Ward, C., Jones, K. W., Coles, R., Rich, L., Knapp, S., & Madsen, R. (2014b). Mentored research in a tribal college setting: The Northern Cheyenne Case. *Journal of Research in Rural Education, 29*(3), 17. Web.

Ward, E. G., Thomas, E. E., & Disch, W. B. (2014c). Mentor service themes emergent in a holistic, undergraduate peer-mentoring experience. *Journal of College Student Development, 55*(6), 563–579. https://doi.org/10.1353/csd.2014.0058.

Ward, L., Siegel, M. J., Davenport, Z., Gardner, J. (2012). *First-generation college students: Understanding and improving the experience from recruitment to commencement* (1. Aufl.; 1st ed.). San Francisco, CA: Jossey-Bass.

Ware, P., & Ramos, J. (2013). First-generation college students: Mentoring through social media. *International Journal of Mentoring and Coaching in Education, 2*(2), 149–162. https://doi.org/10.1108/IJMCE-02-2013-0009.

Warikoo, N., & Carter, P. (2009). Cultural explanations for racial and ethnic stratification in academic achievement: A call for a new and improved theory. *Review of Educational Research, 79*(1), 366–394. https://doi.org/10.3102/0034654308326162.

Wasburn, M. H. (2008). One mentor or two: An instrumental case study of strategic collaboration and peer mentoring. *Journal of the First-Year Experience & Students in Transition, 20*(2), 91.

Waters, T., Marzano, R. J., & McNulty, B. (2003). *Balanced leadership: What 30 years of research tells us about the effect of leadership on student achievement*. Aurora, CO: Mid-continent Research for Education and Learning.

Weddle-West, K., Hagan, W. J., & Norwood, K. M. (2013). Impact of college environments on the spiritual development of African American students. *Journal of College Student Development, 54*(3), 299–314. https://doi.org/10.1353/csd.2013.0050.

Weeks, V. (2014, Apr. 17). Positive self-talk. *Sun Reporter*, pp. L1.

Whalen, D. F., & Shelley, M. C., II. (2010). Academic success for STEM and non-STEM majors. *Journal of STEM Education: Innovations and Research, 11*(1–2), 45.

Wawrzynski, M. R., Jessup-Anger, J. E., & Yao, C. W. (2011). Enhancing undergraduate education: Examining faculty experiences during their first year in a residential college and exploring the implications for student affairs professionals. *The Journal of College and University Student Housing, 38*(1), 56–69.

Whelan, D. L. (2002). Congress passes TEACH act. *School Library Journal, 48*(11), 19.

Wilder, M. (2000). Increasing African American teachers' presence in American schools: Voices of students who care. *Urban Education, 35*(2), 205–220. https://doi.org/10.1177/0042085900352005.

Wilkins, A. C. (2014). Race, age, and identity transformations in the transition from high school to college for black and first-generation white men. *Sociology of Education, 87*(3), 171–187. https://doi.org/10.1177/0038040714537901.

Willcoxson, L., Cotter, J., & Joy, S. (2011). Beyond the first-year experience: The impact on attrition of student experiences throughout undergraduate degree studies in six diverse universities. *Studies in Higher Education, 36*(3), 331–352. https://doi.org/10.1080/03075070903581533.

Williams, M. D., & Dodge, B. J. (1992). Tracking and analyzing learner-computer Interaction. Paper presented at the 1992 National Conference of the Association for Educational Communications and Technology, New Orleans, LA.

Williams, A. P. (2012). *Reentry of substance abusing female ex-offenders from prison to an urban community* (Ph.D. dissertation). Available from ProQuest Dissertations & Theses Full Text. (1009050707).

Williams, P. J. (2011). STEM education: Proceed with caution. *Design and Technology Education: An International Journal*, 16(1), 26–35.

Williams S. R. & Manning, N.F. (2014). *The sophomore transition: Considerations for effective academic advising. Retrieved from NACADA Clearinghouse Resources* web site. http://www.nacada.ksu.edu/Resources/Clearinghouse/View-Articles/The-sophomore-transition-Considerations-for-effective-academic-advising.aspx

Willis, A. I., & Lewis, K. C. (1999). Our known everydayness: Beyond a response to white privilege. *Urban Education, 34*(2), 245–262. https://doi.org/10.1177/0042085999342006.

Wilson, L. V. E., Coulter, B., Lunnen, K., Mallory, J., & Williams, M. (1999). Improving undergraduate student retention. *Allied Academies International Conference Academy of Educational Leadership Proceedings, 4*(2), 36.

Wingert, P. (2010). Minority report. *Newsweek, 155*(9).

Wolff, T. L., Bryant, P. S., & USA Group Noel-Levitz, Iowa City, IA. (1999). *Top ten trends in enrollment management. Synopsis.* Distributed by ERIC Clearinghouse.

Wolniak, G. C. (2016, 2015). Examining STEM Bachelor's degree completion for students with differing propensities at college entry. *Journal of College Student Retention: Research, Theory & Practice, 18*(3), 287-309. https://doi.org/10.1177/1521025115622782.

Wolverton, B. (2007, Dec. 21). Lender agrees to end contracts that allowed it to use sports logos. *The Chronicle of Higher Education*, 19.

Yanico, B. J., & Lu, T. G. C. (2000). A psychometric evaluation of the six-factor self-concept scale in a sample of Racial/Ethnic minority women. *Educational and Psychological Measurement, 60*(1), 86–99. https://doi.org/10.1177/00131640021970385.

Young, L., & Papinczak, T. (2013). Strategies for sustaining quality in PBL facilitation for large student cohorts. *Advances in Health Sciences Education, 18*(4), 825–833. https://doi.org/10.1007/s10459-012-9387-7.

Yun, J. H., Baldi, B., & Sorcinelli, M. D. (2016). Mutual mentoring for early-career and underrepresented faculty: Model, research, and practice. *Innovative Higher Education, 41*(5), 441–451. https://doi.org/10.1007/s10755-016-9359-6.

Yusoff, M. S. B., Rahim, A. F. A., Babba, A. A., Iismail, S. B., & Esa, A. R. (2012). A study of psychological distress in two cohorts of first-year medical students that underwent different admission selection processes. *Malaysian Journal of Medical Sciences, 19*(3), 29–35.

Zawistowska, A. (2017). Is entering STEM socially contagious? Contextual factors in women's educational decisions. *Polish Sociological Review, 197*, 51.

Zevallos, A. L., & Washburn, M. (2014). Creating a culture of student success: The SEEK scholars peer mentoring program. *About Campus, 18*(6), 25–29. https://doi.org/10.1002/abc.21141.

Zhan, W. (2014). Research experience for undergraduate students and its impact on STEM education. *Journal of STEM Education: Innovations and Research, 15*(1), 32.

Zhang, L., & Barnett, M. (2015, 2014). How high school students envision their STEM career pathways. *Cultural Studies of Science Education, 10*(3), 637-656. https://doi.org/10.1007/s11422-013-9557-9.

Zhang, P., & Smith, W. L. (2011). From high school to college: The transition experiences of black and white students. *Journal of Black Studies, 42*(5), 828–845. https://doi.org/10.1177/0021934710376171.

Zoller, U. (2011). Science and technology education in the STES context in primary schools: What should it take? *Journal of Science Education and Technology, 20*(5), 444–453. https://doi.org/10.1007/s10956-011-9306-3.

Zoller, U. (2015). Research-based transformative Science/STEM/STES/STESEP education for "sustainability thinking": From teaching to "know" to learning to "think". *Sustainability, 7*(4), 4474–4491. https://doi.org/10.3390/su7044474.

The manufacturer's authorised representative in the EU is Springer Nature Customer Service Centre GmbH, Europaplatz 3, 69115 Heidelberg, Germany. If you have any concerns regarding our products, please contact ProductSafety@springernature.com

Printed and bound by CPI Group (UK) Ltd, Croydon, CR0 4YY
30/09/2026
02225709-0002